Software Engineering Design Knowledge Areas

Volume 2

The Engineering of Software Projects

The Engineering of Software Projects

Software Engineering Knowledge Areas

These volumes support the IEEE *Guide to Software Engineering Body of Knowledge (SWEBOK)* and the IEEE Computer Society Professional Software Engineering Master Certification.

This is a Work-in-Progress; each of these volumes is not currently available, but is expected to be published in the coming year.

— *Richard Hall Thayer*

Volume 1   —   Software Engineering Requirements

Volume 2   —   Software Engineering Design

Volume 3   —   Software Engineering Construction

Volume 4   —   Software Engineering Testing

Volume 5   —   Software Engineering Maintenance

Volume 6   —   Software Engineering Configuration Management

Volume 7   —   Software Engineering Management

Volume 8   —   Software Engineering Processes

Volume 9   —   Software Engineering Models and Methods

Volume 10   —   Software Engineering Quality Assurances

Volume 11   —   Software Engineering Economics

Volume 12   —   Software Engineering Project Management

These volumes will be published and sold through Amazon Books.

Software
Knowledge Engineering
Design Areas

Volume 2
The Engineering of Software Projects

Richard Hall Thayer, PhD, CSDP

Contributing authors:

David Budgen, Professor, University of Durham, U.K.

Hassan Gomaa, PhD, George Mason University

Software Management Training
Carmichael, California
2017

Copyright Notices

Table of Contents

Honorary Foreword..vii

Preface...ix

Acknowledgments...xi

Chapter 1: Software Engineering Design Fundamentals 1

    1. Software Engineering Design Fundamentals................................. 4

    2. Key Elements in Software Design 9

    3. Software Architecture, Structure and View Points 12

    4. Detailed Design 20

    5. User Interface Design 23

    6. Software Design Notations...33

    7. Software Design Strategies and Methods 35

    8. Design Tools 46

    Appendix A: Architectural styles..47

    REFERENCES 53

Chapter 2: The Nature of Software Design..57

    David Budgen, University of Durham, U.K.

    1. The Nature of Software Design 57

    2. Software Architecture 61

    3. Describing Designs 64

    4. Organizing the Design Process 71

    5. Discussion..82

    REFERENCES..82

Chapter 3: Model-Based Software Design for Concurrent and Real-Time Systems 87

    Hassan Gomaa, PhD, George Mason University

    1. Introduction..87

    2. Concurrent Processing Concepts 88

    3. Run-Time Support for Concurrent Tasks 88

    4. Survey of Design Methods for Concurrent and Real-Time Systems 88

    5. Software Design Method for Concurrent/Real-Time Embedded Systems 90

    6. Software Architectural Patterns for Real-Time Control 101

    7. Performance Analysis of Real-Time Designs 102

    8. Real-Time Embedded Software Product Line Design 102

    9. Conclusions,... 103

    REFERENCES...104

Chapter 4: Software Design Documentation Standard..........................109

    1. Outline of a Software Design Description..................................110

    2. General Design Considerations 111

    3. Architectural Design Considerations...112

    4. Detailed Design Considerations...114

Chapter 5: Software Design Exercises 119

INDEX...125

A Partial List of General Abbreviations
(One-of-a-kind abbreviations are identified in place)

a.k.a.	—	also known as
API	—	application programming interface
a.s.a.p.	—	as soon as possible
ConOps	—	concept of operations (document)
CSCP	—	Computer Society Certificates of Proficiency
DFD	—	data flow diagram
FSM	—	finite state machines
HCI	—	human computer interface
HW	—	hardware
I/O	—	input/output
IV&V	—	independent verification and validation
KA	—	knowledge area
PSEM	—	Professional Software Engineering Master (Certification)
SCM	—	software configuration management
SED	—	software engineering design
SEM	—	software engineering management
SEPM	—	software engineering project management
SET	—	software engineering testing
SQA	—	software quality assurance
SW	—	Software
SRE	—	software requirements engineering
SWE	—	software engineering
SWEBOK	—	(Guide to the) Software Engineering Body of Knowledge
TBD	—	to be determined/done
V&V	—	verification and validation

Honorary Foreword

To explain the origin of the term "software engineering," the following story is offered.[1]

In the mid-1960s, there was increasing concern in scientific quarters of the Western world that the tempestuous development of computer hardware was not matched by appropriate progress in software development. The software situation looked to be more turbulent. Operating systems had just become the latest rage, but they showed unexpected weaknesses. The uneasiness had been articulated in the NATO Science Committee by its U.S. representative, Dr. I.I. Rabi, the Nobel laureate and famous, as well as influential, physicist. In 1967, the Science Committee set up the Study Group on Computer Science, with members from several countries, to analyze the situation.

The German authorities nominated me for this team. The study group was given the task of "assessing the entire field of computer science," with particular elaboration on the Science Committee's consideration of organizing a conference and, perhaps later, setting up an International Institute of Computer Science.

The study group, concentrating its deliberations on actions that would merit an international rather than a national effort, discussed all sorts of promising scientific projects. However, it was rather inconclusive on the relation of these themes to the critical observations mentioned above, which had guided the Science Committee in creating the study group,

Perhaps not all members of the study group had been properly informed about the rationale for its existence. In a sudden mood of anger, I remarked, "The whole trouble comes from the fact that there is so much tinkering with software. It is not made in a clean fabrication process." When I found out that this remark was shocking to some of my scientific colleagues, I elaborated on the idea with the provocative saying, "What we need is *software engineering*."

This remark caused the expression "software engineering," which seemed to some to be a contradiction in terms, to be stuck in the minds of the members of the group. In the end, in late 1967, the study group recommended that we hold a Working Conference on Software Engineering, and I was made chairman. I not only had the task of organizing the meeting (which was held from October 7 to October 10, 1968, in Garmisch, Germany), but I had to set up a scientific program for a subject that was suddenly defined by my provocative remark.

---

1. Dr. Bauer originally wrote this paper as an introduction to a 1993 IEEE tutorial: *Software Engineering: A European Perspective*, R.H. Thayer, and A.D. McGettrick, eds., IEEE Computer Society Press, Los Alamitos, CA, 1993.

I enjoyed the help of my co-chairmen, L. Bolliet from France and H. J. Helms from Denmark. In addition, I had the invaluable practical support of the programcommittee members, A.J. Perlis and B. Randell in the section on design, P. Naur and J.N. Buxton in the section on production, and K. Samuelson, B. Galler, and D. Gries in the section on service.

Among the 50 or so participants, E.W. Dijkstra was dominant. Not only did he make cynical remarks, like, "The dissemination of error-loaded software is frightening," and "It is not clear that the people who manufacture software are to be blamed. I think manufacturers deserve better, more understanding users," but he also said, at this early date, "Whether the correctness of a piece of software can be guaranteed or not depends greatly on the structure of the thing made." He had very fittingly named his paper, "Complexity Controlled by Hierarchical Ordering of Function and Variability," introducing a theme that followed his life over the next 20 years. Some of his words have become proverbs in computing, like, "Testing is a very inefficient way of convincing oneself of the correctness of a program."

With the wide distribution of the reports from the Garmisch Conference and in a follow-up conference in Rome, from October 27 to October 31, 1969, it happened that not only the term "software engineering" but also the idea behind this term became fashionable. Chairs were created, institutes were established (although the one that the NATO Science Committee had proposed did not come about because of reluctance on the part of Great Britain to have it organized on the European continent), and a great number of conferences were held.

The tutorial nature of the papers in this book is intended to offer readers an easy introduction to the topics and indeed to the attempts that have been made in recent years to provide them with the *tools,* both in a handcraft and an intellectual sense, which allow them now to honestly call themselves *software engineers.*

Friedrich L. Bauer, PhD
Professor Emeritus

Technische Universität München (TUM)
Germany

> P.S. In 1989, I met Dr. Friedrich L. Bauer, Professor Emeritus, Universität München, when I was giving a software engineering seminar in Munich for the IEEE. Later, Professor Bauer provided me with the story describing how he came to name what we now call *software engineering* (I reprinted the story as an Honorary Foreword by Professor Bauer). I thought some of the readers, especially the younger ones, might be interested in the history of the naming of software engineering. Professor Bauer passed awayl in 2015 at the age of 90.   RHT

Preface

Software design engineering is the process of determining what is to be produced in a software system. It has the widely recognized goal of determining the needs for, and the intended external behavior of, a system design.

University students as well as candidates for the IEEE Computer Society Certificate of Proficiency exam in *software design* need to focus on the following five subareas of the design knowledge area [https://www.computer.org/web /education/re-design]:

1. *The successful application of essential design principles and methods for the overall software design process.*

2. *The knowledge and skills necessary to apply the concepts of concurrency, data persistence, error handling, and security to a typical software design project, understanding the essential elements of software structure and architecture in terms of styles, patterns, and families of programs, and frameworks within software development projects.*

3. *Having a command of the key principles involved in the development of user interface design to a software development project, including essential principles, interaction modalities, information presentation, and the UI design process, understanding the appropriate application of quality analysis and evaluation principles including quality attributes, analysis and evaluation techniques, and quality measures.*

4. *The knowledge and ability to employ design notations in terms of structure and behavioral descriptions in the software design process.*

5. *The knowledge and ability to employ function, object, data-structure, and component-based design methodologies in a typical software design project.*

Our book makes maximum use of SWEBOK—a very impressive document—and should be read by anybody studying software engineering.

To accommodate both groups—university students as well as candidates for the IEEE Computer Society Certificate of Proficiency exam in *software design*—a software engineering principle that is not included in SWEBOK 2014 and is <u>not</u> likely to produce an exam question is marked with the following statement: *"Note: SWEBOK does not include (to be filled in) in the SWEBOK guide.* The certificate candidate is free to skip this entry. The university student should probably study this entry.

This book uses the outline (paragraph headings) expressed in *Software Design,* (Chapter 2), of *Guide to Software Engineering Body of Knowledge* (SWEBOK) IEEE Computer Society, 2014, as the outline for this chapter.

This small book is divided into five parts.

1. Chapter one presents an analysis of the appropriate software engineering knowledge areas (KAs) followed by an explanation of the material related to the topics contained in the SWEBOK.

2. Two additional articles from highly regarded technical sources are:

   a. *The Nature of Software Design,* Professor David Budgen, University of Durham, U.K.

   b. *Software Design for Concurrent and Real-Time Systems, Dr.* Hassam Gomaa, George Mason University.

3. The third part of the book includes a software design documentation standard (SDS). This standard is modeled after *IEEE Standard 1016-1998,* "IEEE Recommended Practice for a Software Design Description." The purpose of this document is to: (1) aid the IEEE *Software Design Certificate of Proficiency Exam* candidates in taking and passing the software design exam, (2) provide an example of a software design standard to demonstrate the usefulness of an IEEE software engineering standard, and to (3) provide a templet for university students to use in writing an SDS for a classroom software engineering project.

   This classroom standard should not be used to satisfy a commercial software engineering contract. Nevertheless, it can be used as an educational tool for university students and Software Design Certification exam takers.

4. Twenty sample exam questions to aid both the exam takers and university students are included in Part 4.

5. Part 5 contains an index referencing key concepts and thinkers in the field of Software Design..

Richard Hall Thayer, PhD, CSDP
Life Fellow of the IEEE
Member of the IEEE Computer Society Golden Core
Emeritus Professor of Software Engineering,
   Sacramento State University, California

Acknowledgments

No successful endeavor has ever been undertaken by one person alone. I would like to thank the following people and organizations who supported me in this effort.

I first want to thank my wife Mildred for her high degree of tolerance in putting up with my working seven days a week on this manuscript. Without her encouragement and support, this book could never have been completed.

I also want to thank Ellen Sander who performed copyediting, Jon Digerness of North Coast Graphics for providing me with the comic illustrations, and Jim Tozza for giving me hardware and software support to keep my computer equipment running.

In addition, I want to thank Steve Tockey for providing me with help about the Computer Society exam specifications in order to maximize the usefulness of our software engineering textbook and SWE guidebook, and Melville (Mel) Piercey of Copy Plus for providing cover artwork and designing and drawing the engineering chapter graphics.

Finally, I want to thank our little dog Maxwell (a.k.a. Max, Maxcito, or Speedy) who kept me company in the evening hours when everybody else had gone to bed.

A happy Max says that:

This is a Terrrrrrrrific Book. I chewed on a copy,
and it was tasty.

A Note to Our Readers

One of the advantages of using a "print-on-demand" (POD) publishing service is the ability to easily make changes to the manuscript when errors or improvements are identified.

The authors encourage you to identify and send potential errors or suggested improvements to the e-mail address listed below. I do not guarantee to make all the changes identified, but I do promise to review and seriously consider all recommendations.

Disclaimer

While I have more than 50 years of software engineering experience, including university teaching, I am not a technical expert in every component of software engineering. To make up for this shortcoming, I have made extensive use of material written by subject matter experts and papers (many posted on the web) as source documents.

Every effort has been made to make this software engineering reference as complete and accurate as possible. However, I can make no representation or warranties with respect to accuracy or completeness of the contents of this book and specifically disclaim any implied warrantee of merchantability or fitness for a particular purpose. The advice and strategies contained herein may not be suitable for your situation. If in doubt, you should consult with a professional software engineer. Where appropriate, neither I nor the printer we be liable for the loss of profit or other commercial damages, including, but not limited to special, incidental, consequential, or other damages [IEEE Press disclaimer].

Please keep me posted.

Richard Hall Thayer, PhD, CSDP
thayer@csus.edu

Chapter 1
Software Engineering Design Fundamentals

This chapter is a textbook and study guide to introduce the principles and some of the problems of software engineering design (SED). This book can be used either in a university course in software design or as a study guide to aid individual software engineers in passing the IEEE Professional Software Engineering Master (PSEM) Certification exams in software design.

This chapter examines the role and context of the design activity as a form of problem-solving process, describes how this is supported by current design methods and methodologies, and considers the strategies, strengths, limitations, and main domains of the application of these methods.

Software (SW) design plays an important role in developing software. It allows software engineers to produce various models that form a blueprint of the solution to be implemented. We can analyze and evaluate these models to determine whether they will allow SW engineers to fulfill the various requirements specifications. Design develops a human computer interface (HCI) that assists users in making maximum use of the software system. Finally, we can use the models to plan subsequent development activities, in addition to using them as input and the starting point of construction and testing.

An Interface Problem

INTRODUCTION

Design is defined as both the process of defining the architecture, components, interfaces, and other characteristics of a system or component and the result of that process (a software design description) [SWEBOK 2004]. Viewed as a process, SED is the software engineering life cycle activity in which software requirements are analyzed in order to produce a description of the software's internal structure that will serve as the basis for its construction. Software design includes *architectural design* (sometimes-called "high-level design" or incorrectly called "preliminary design") and *detailed design* (sometimes incorrectly called "critical design") that fits between software requirements analysis and software engineering construction.

Architectural design is the process of defining a collection of hardware and software components and their interfaces to establish a framework for the development of a software system. *Detailed design* defines the components that fit into the architectural design framework. The design KA is divided into eight sub-areas [SWEBOK 2004]:

(1) **Software design fundamentals** form an underlying basis to the understanding of the role and scope of SED. These are general software concepts, the context of SED, the SED process, and enabling techniques.

(2) **Key issues in SED** include concurrency, control, the handling of events, and distribution of components. Included are human error, exception handling and fault tolerance, interaction, presentation, and data persistence.

(3) **Software structure and architecture** includes topics on architectural structures and viewpoints, architectural styles, design patterns, and finally, families of programs and frameworks.

(4) **User interface design** should ensure that interaction between the human and the machine provides for effective operation and control of the computing machine.

(5) **Software design quality analysis and evaluation** presents the topics specifically related to SED. These aspects are quality attributes, quality analysis, evaluation techniques, and measures.

(6) **Software design notations** are divided into structural and behavioral descriptions.

(7) **Software design strategies and methods** describe general strategies, followed by function-oriented design methods, object-oriented design methods, data-structure-centered design, component-based design, and others.

(8) **Software design tools** can be used to support the creation of SED artifacts during software development.

SOFTWARE DESIGN KNOWLEDGE AREAS

Software design is a process of defining the architecture, components, interfaces, and other characteristics of a system or component and planning for a software solution. Figure 1 provides a top-level decomposition and breakdown of the SED knowledge area (KA). Figure 2 shows a sequence of software development phases, their relationships with each other, and the major software product resulting from each phase. The phases and products associated with SED are marked with a "star."

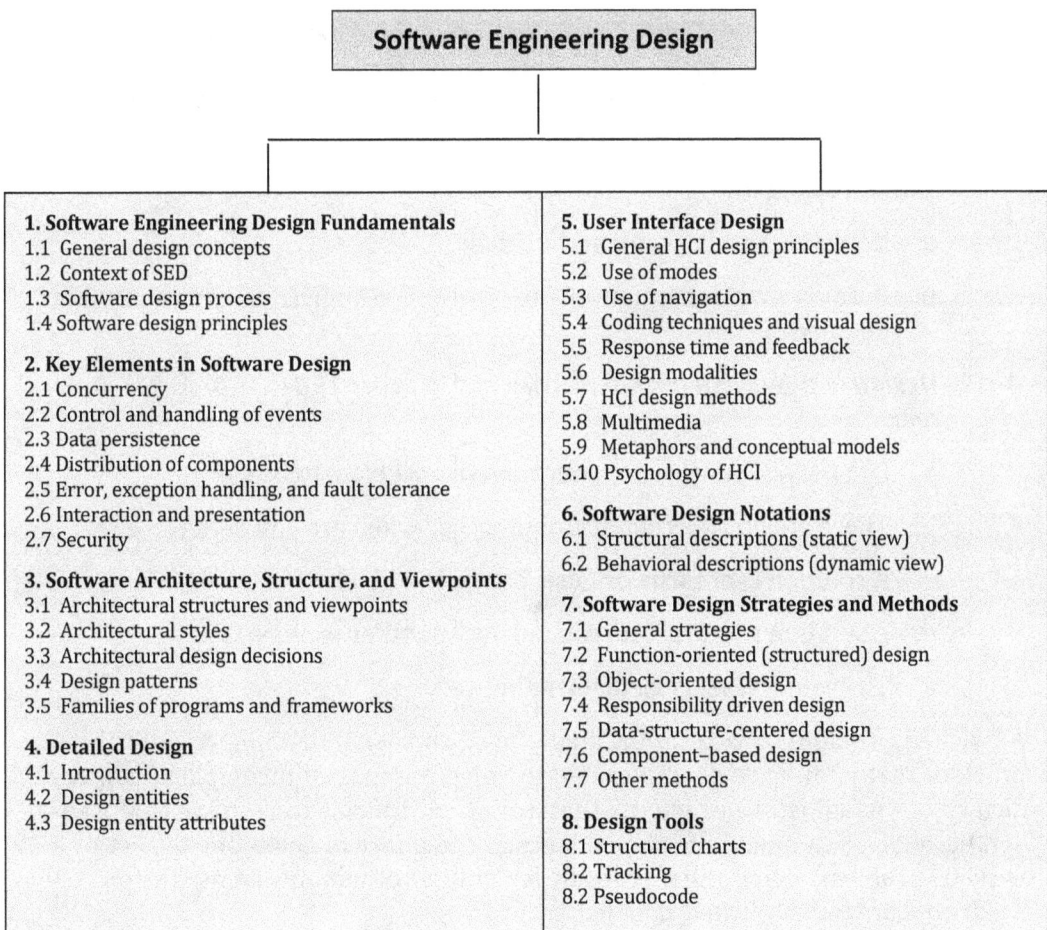

Software Engineering Design

1. Software Engineering Design Fundamentals
1.1 General design concepts
1.2 Context of SED
1.3 Software design process
1.4 Software design principles

2. Key Elements in Software Design
2.1 Concurrency
2.2 Control and handling of events
2.3 Data persistence
2.4 Distribution of components
2.5 Error, exception handling, and fault tolerance
2.6 Interaction and presentation
2.7 Security

3. Software Architecture, Structure, and Viewpoints
3.1 Architectural structures and viewpoints
3.2 Architectural styles
3.3 Architectural design decisions
3.4 Design patterns
3.5 Families of programs and frameworks

4. Detailed Design
4.1 Introduction
4.2 Design entities
4.3 Design entity attributes

5. User Interface Design
5.1 General HCI design principles
5.2 Use of modes
5.3 Use of navigation
5.4 Coding techniques and visual design
5.5 Response time and feedback
5.6 Design modalities
5.7 HCI design methods
5.8 Multimedia
5.9 Metaphors and conceptual models
5.10 Psychology of HCI

6. Software Design Notations
6.1 Structural descriptions (static view)
6.2 Behavioral descriptions (dynamic view)

7. Software Design Strategies and Methods
7.1 General strategies
7.2 Function-oriented (structured) design
7.3 Object-oriented design
7.4 Responsibility driven design
7.5 Data-structure-centered design
7.6 Component–based design
7.7 Other methods

8. Design Tools
8.1 Structured charts
8.2 Tracking
8.2 Pseudocode

Figure 1: Hierarchy listing of the SWE design KAs

1. Software Engineering Design Fundamentals

The concepts, notions, and terminology introduced in this chapter form an underlying basis for understanding the role and scope of SED.

1.1 General design concepts.

Software design is a problem-solving process in which the designer applies knowledge and experience to define and describe a solution to a technical problem described in the software requirements specification. Software design bridges the gap between software requirements and software code.

The term "software design" includes all that is involved between the conception of the desired software through to the final manifestation of the software, ideally in a planned and structured process [http://en.wikipedia.org/wiki/Software_development].

(1) **Software design**:

    a.   Describes the software architecture.

    b.   Produces a description of the software's internal construction. i.e., detailed design.

(2) **Design is a "wicked problem"** [Riel and Webber 1984]. In a *wicked problem*:

    a.   It is not always easy to determine the problem to be solved.

    b.   There are no rules about stopping, i.e., when are you done?

    c.   A solution is not true or false, but good or bad.

    d.   A solution to a problem can uncover another problem.

    e.   The solution set is virtually unlimited.

A "wicked problem" is a term originally used in social planning to describe a problem that is difficult or impossible to solve because of incomplete, contradictory or changing requirements that are often difficult to recognize. Moreover, because of complex interdependencies, the effort to solve one aspect of a wicked problem may either reveal or create other problems [http://en.wikipedia.org/wiki/Wicked_problem].

1.2 Context of SED.

To understand the role of SED, it is important to understand the context in which it fits—the software engineering life cycle (*see Figure 2*). Thus, it is important to understand the major characteristics of software requirements analysis versus software engineering design versus software engineering construction versus software testing [SWEBOK 2004].

(1) **Software requirements** — A subfield of software engineering that uses elicitation, analysis, specification, and validation of requirements for software and project planning [http://en.wikipedia.org/wiki/Software _requirements].

(2) **Software design** — A process of problem solving and project planning for a software solution. After the purpose and specifications of software are determined, software developers will either design or employ designers to develop a technical plan for a solution to the SW requirements listed in the SRS. [http://en.wikipedia.org/wiki/Software_design].

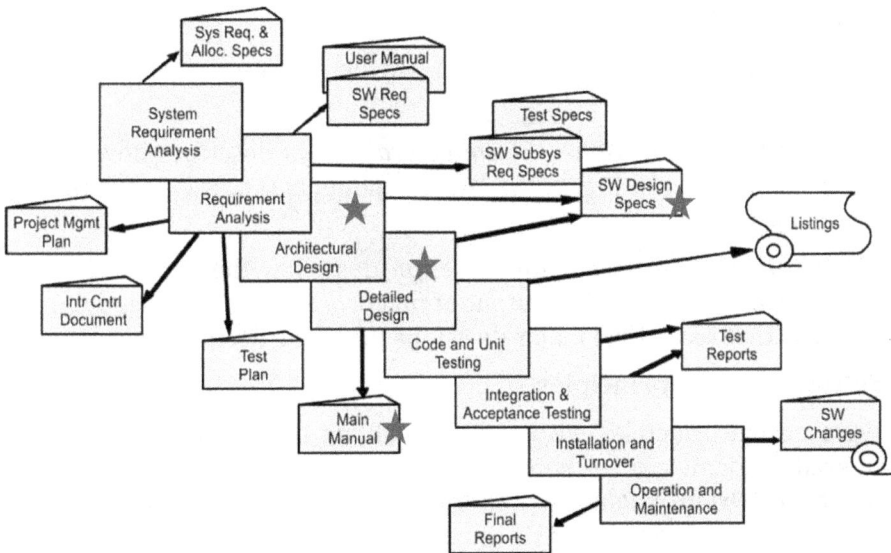

Figure 2: Sequence of software development phases

(3) **Software construction** — Software construction (a.k.a. software development, application development, designing software, software application development, enterprise application development, or platform development) is the construction (i.e., coding and testing) of a software product.

The term "software construction" may also be used to refer to the activity of *computer programming*, which is the process of writing and maintaining the source code. Nevertheless, in a broader sense, the term includes everything that is involved between the conception of the desired software through to the final manifestation of the software, ideally in a planned and structured process [http://en.wikipedia.org/wiki/Soft ware_development].

(4) **Software testing** — Software testing is an investigation into the completed, or partially completed, software system to provide stakeholders with information about the quality of the product or service under test. Software testing is a review or inspection (i.e., called *static testing*) of the developed SW system to determine whether there are errors in the system. It is also the exercising of operating software systems (called *dynamic testing*) to look for errors in the SW system. However, SW testing cannot "prove" that the SW does <u>not</u> have errors [https://en.wikipedia .org/wiki/Software_testing].

1.3 Software design process.

Software design is generally considered to be a two-step process [SWEBOK 2004].

(1) **Architectural design** — *Architectural design* describes how software is decomposed and organized into components (i.e., the software architecture).

(2) **Detailed design** — *Detailed design* describes the specific behavior of these components. The output of this process is a set of models and artifacts that record the major decisions that have been made.

1.4 Software design principles.

Software design principles, also called "enabling techniques," are key notions considered fundamental to many different SED approaches and concepts. According to the *Oxford English Dictionary*, a "principle" is "a basic truth or a general law . . . that is used as a basis of reasoning or a guide to action." Some SED principles or enabling techniques are [SWEBOK 2004]:

(1) **Abstraction** — *Abstraction* is "the process of forgetting information so that things that are different can be treated as if they were the same" [Liskov & Guttag 2001].

In the context of SED, two key abstraction mechanisms are specification and parameterization [http://en.wikipedia.org/wiki_Abstraction].

a. **Abstraction by specification** — A well-designed specification removes unnecessary detail about the actual type or value being specified.

b. **Abstraction by parameterization** — Rather than write code that mentions specific values on which computation is to occur, we create functions. Functions describe a computation that works on all acceptable values of the appropriate types. Thus, the detail of what specific values are to be used is removed. Parameterized types are another example of abstraction by parameterization, although the parameters are types not values.

Abstraction makes the job of all parties simpler and makes the code more extensible and maintainable [http://www.cs.cornell_edu /course/cs312/2007sp/lectures/lec06.html].

This principle is a.k.a. *information hiding (see Paragraph 7.1(4).*

(2) **Coupling** — *Coupling* is the measure of the strength of association between software components. It is a way of evaluating the partitioning of a system. The developer should always strive for loose (low) coupling, for example: (ordered best to worst) [Christensen 2005]:

a. **Data coupling** — Communicates by parameters.

b. **Stamp coupling** — Refers to the same data structure.

c. **Control coupling** — Control information is passed.

d. **Common coupling** — Refers to the same global data.

e. **Contents coupling** — Refers to the insides of another module.

Figure 3 provides a graphic illustration of coupling. The top illustration depicts contents coupling. The lower diagram illustrates data coupling. Contents couplings should be avoided with a software design.

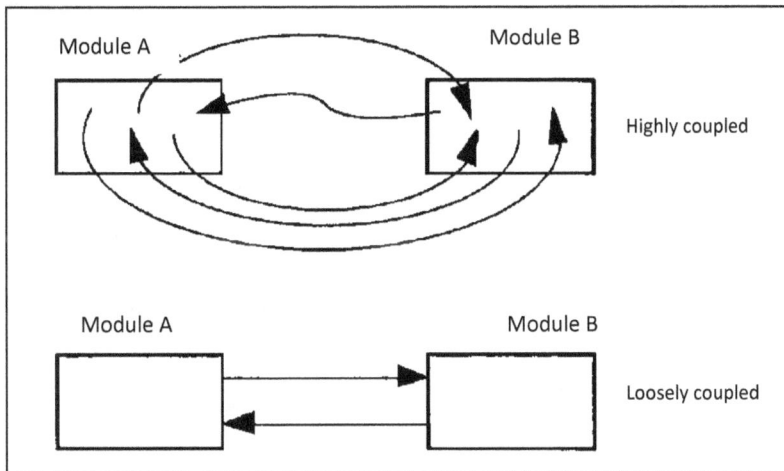

Figure 3: Illustration of coupling

(3) **Cohesion (a.k.a. binding)** — *Cohesion* is a way of measuring the internal connectivity of a system. Always strive for strong cohesion: for example, functional, sequential, and communication binding (*see Figure 4*). Measures of binding (best to worst) are as follows [Christensen 2005]:

a. **Functional binding** — Elements related to a single function.

b. **Sequential binding** — One output is input to another.

c. **Communication binding** — The same input or output data.

d. **Temporal binding** — Related by time.

e. **Procedural binding** — Related by process.

f. **Logical binding** — Logical relationship between elements.

g. **Coincidental binding** — No meaningful relationship.

Binding is a way of measuring the cohesiveness of a system. Binding is concerned only with the internal commonality of a single module being built.

Figure 4 illustrates two examples of binding. The top binding is logical binding, in which all elements have a logical connection. The bottom binding is functional binding, in which the internal functions are grouped about a single function. The top three bindings—functional, sequential, and communication—are considered the strongest internal bindings that represent the goal of modular design.

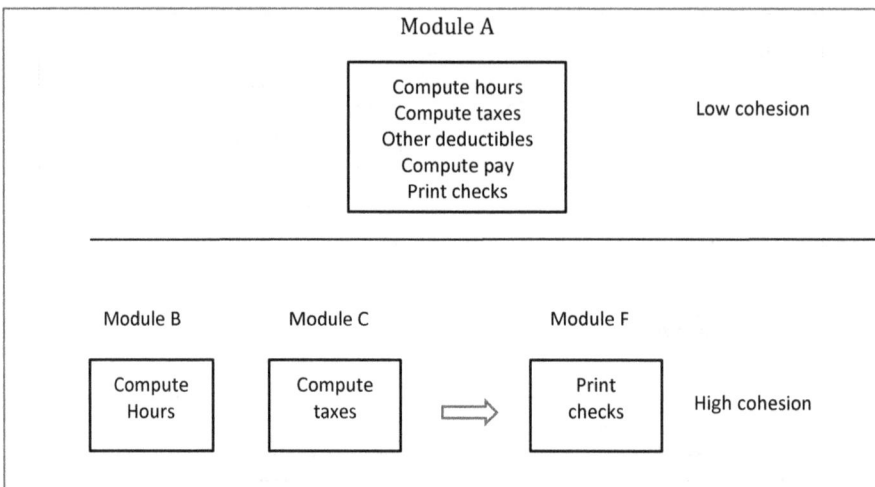

Figure 4: Illustration of cohesion/binding

(4) ***Decomposition and modularization*** — This principle refers to the *decomposition* and *modularization* of a large element of software into a number of smaller independent elements, usually with the goal of placing different functionalities with different components.

(5) ***Encapsulation/Information hiding*** — *Encapsulation* or *information hiding* means grouping and packaging the components and internal details of an abstraction and making those details inaccessible to individuals who do not "have a need to know," for example customers of potential users.

(6) ***Separation of interface and implementation*** — *Separation of interface* and *implementation* involves defining a component by specifying a public interface and separating it from the details describing how the component is realized.

(7) ***Sufficiency, completeness, and primitiveness*** — Achieving *sufficiency, completeness*, and *primitiveness* means ensuring that a software component captures all the important characteristics of an abstraction and *nothing more*. (In Scotland this is called being "parsimonious.")

(8) ***Separation of concerns*** — *Separation of concerns* is a design principle used for separating a computer program into distinct sections, such that each section addresses a separate concern. A concern is an information set that affects the code of a computer program. A program that embodies this principle is called a "modular program" [http://en.wikipedia.org /wiki/Separation_of_concerns].

2. Key Elements in Software Design

A number of key elements must be dealt with when designing software. Some are quality concerns that all software must address (for example, performance, security, reliability, usability, and so forth). Another important issue is how to decompose, organize and package software components. In contrast, there are issues that deal with some aspect of software's behavior that are not in the application domain, but address the supporting domains [Bosch 2000].

Such issues, which often crosscut the system's functionality, have been referred to as some features that tend not to be units of software's functional decomposition, but rather to be properties that affect the performance or semantics of the components in systemic ways [Kiczales et al. 1997]. The following paragraphs contain a number of these key elements—listed in alphabetical order [SWEBOK 2004].

2.1 Concurrency.

Because computations in a concurrent system can interact with each other while they are executing, the number of possible execution paths in the system can be extremely large, and the resulting outcome can be indeterminate. Concurrent use of shared resources can be a source of indeterminacy leading to issues such as *deadlock* and *starvation*.

(1) **Deadlock** — A *deadlock* occurs when two competing actions wait for the other to finish, and thus neither ever does. Deadlock is a common problem in multiprocessing systems, parallel computing, and distributed systems, where software and hardware locks are used to handle shared resources and implement process synchronization [https://en.wikipedia .org/wiki/Deadlock].

(2) **Starvation** — *Starvation* is a problem encountered where a process is perpetually denied necessary resources to process its work. Starvation may be caused by errors in a scheduling or mutual exclusion algorithm, but can also be caused by resource leaks, and can be intentionally caused via a denial-of-service attack [https://en.wikipedia.org/wiki/Starvation _%28computer_science%29].

The design of *concurrent systems* often entails finding reliable techniques for coordinating execution, data exchange, memory allocation, and execution scheduling to minimize response time and maximize throughput [http://en .wikepedia.org/wikiConcurrency_(computer_science)].

2.2 Control and handling of events.

Event-driven systems are typically used when some asynchronous external activity needs to be handled by a program as, for example, when a user presses a button on the mouse [https://en.wikipedia.org/wiki/Event-driven_architecture].

(1) *Events* are typically used in user interfaces where actions in the outside world (for example, mouse clicks, window resizing, keyboard presses, initiating the autopilot, light gun fired, messages from other programs, etc.) are handled by the program as a series of events. Programs written for many windowing environments consist predominantly of event handlers.

(2) An *event-driven system* typically runs an event loop that keeps waiting for such activities (e.g., input from devices or internal alarms). When one of these occurs, it collects data about the event and fires it; in other words, it dispatches the event to the proper event handler software that will address the activity.

(3) A program can choose to *ignore events* and there may be programs to listen for a particular event. At a minimum, the data associated with an event specifies the type of event it is.

2.3 Data persistence.

A *persistent data structure* is a data structure that always preserves the previous version of itself when it is modified. Such data structures are effectively immutable, as their operations do not (visibly) update the structure in-place but instead always yield a new, updated structure. A data structure is *partially persistent* if all versions can be accessed but only the newest version can be modified. The data structure is *fully persistent* if every version can be both accessed and modified.

2.4 Distribution of components.

Distribution of components concerns how to distribute the software across the hardware, how the components communicate, and how middleware can be used to deal with heterogeneous software. Virtually all large computer-based systems are now distributed systems.

2.5 Error, exception handling, and fault tolerance.

(1) **Error handling** — Error handling refers to the programming practice of anticipating and coding for error conditions that may arise when the program runs. Errors in general occur in three forms [http://www.cpearson.com/excel/errorhandling.htm]:

   a. **Compiler errors** such as undeclared variables that prevent the code from compiling.

   b. **User data-entry errors** such as a user entering a negative value where only a positive number is acceptable.

   c. **Run-time errors** that occur when the program cannot correctly execute a program statement.

(2) **Exception handling** — Exception handling is a programming language construct or computer hardware mechanism designed to handle unusual occurrences or exceptions (i.e., special conditions that change the normal flow of program execution). Programming languages differ considerably in their support for exception handling (as distinct from error checking, which is normal program flow that codes for responses to adverse contingencies such as invalid state changes or the unsuccessful termination of invoked operations) [http://en.wikipedia.org/wiki/Exception_handling].

(3) **Fault tolerance** — *Fault-tolerance* (a.k.a. *graceful degradation*) is the property that enables a computer system to continue operating properly in the event of a component failure. If the system operating quality de-

creases at all, the decrease is proportional to the severity of the failure, as compared to a naïvely-designed system in which even a small failure can cause total breakdown [http://en.wikipedia.org/wiki /Fault-tolerantsystem].

2.6 Interaction and presentation.

These are methods used to structure and organize *interactions* between users and the *presentation* of information. An effective *presentation* makes the best *use* of the relationship *between* the presenter and the audience. Many of the best presenters can establish an interaction between themselves and their audience.

2.7 Security.

Software Security Assurance (SSA) is the process of ensuring that software is designed to operate at a level of security that is consistent with the potential harm that could result from the loss, inaccuracy, alteration, unavailability, or misuse of the data and resources that it controls and that protects the SW system [http://en.wikipedia.org/wiki/Software_securit_asinsurance].

3. Software Architecture, Structure, and Viewpoints

Software architecture refers to the high-level structures of a software system, the discipline of creating such structures, and the documentation of these structures. These structures are needed to reason through the components of the software system. Each structure comprises software elements, relations among them, and properties of both elements and relations '[SWEBOK 2004].

Each software engineering system is comprised of separate, semi-autonomous components, called "configuration items (CI)." This collection of configuration items is termed the "architecture" of the system. Each CI is comprised of software design entries (a.k.a. software modules). Each entity has a name and is described independently (*See Paragraph 4.3.1*). Interfaces between configuration items need to be kept to a minimum.

The *architecture* of a software system is a metaphor, analogous to the architecture of a building. Software architecture is about making structural choices that are difficult to change after implemented. Software architecture includes specific structural options taken from software design possibilities.

For example, the systems that controlled the space shuttle launch vehicle had the requirement of being very fast and very reliable. Therefore, an appropriate real-time computing language would need to be chosen. Additionally, to satisfy the need for reliability, the choice could be made to have multiple redundant and independently produced copies of the program, and to run these copies on independent hardware while crosschecking results [https://en.wikipedia.org /wiki/Software_system].

In its strictest sense, *software architecture* is a description of the subsystems and components of a software system and the relationships between them. Architecture thus attempts to define the internal *structure* of software (according to the *Oxford English Dictionary*) as, "the way in which something is constructed or organized." During the mid-1990s, however, software architecture started to emerge as a broader discipline involving the study of SW structures and architectures in a more general way.

This gave rise to a number of interesting ideas about SED at different levels of abstraction. Some of these concepts can be useful during the architectural design of specific software (for example, architectural style), as well as during its detailed design (for example, lower-level design patterns). However, they can also be useful for designing generic systems, and leading to the design of families of programs (also known as product lines). Interestingly, most of these concepts can be seen as attempts to describe, and thus reuse generic design knowledge [SWEBOK 2004].

3.1 Architectural structures and viewpoints.

Different high-level facets of an SED can and should be described and documented. These facets are often called "views." "A view represents a partial aspect of a software architecture that shows specific properties of a software system" [Buschmann 1996]. These distinct views pertain to specific issues associated with SED—for example, satisfying the logical view versus the process view, and the physical view versus the development view.

In summary, an SED is a multifaceted artifact produced by the design process and generally composed of relatively independent views [SWEBOK 2004].

A *view* can consist of a representative set of elements and their relationships. A *view* is the representation of a *structure*. A modular view is the representation of a modular structure. A view is represented by one of a number of different notations selected by the customers and/or developers.

Each module represents and implements some part of the system requirement. Each requirement is allocated to a top-level module. Lower-level modules can be developed through a series of decompositions—starting with top-level system requirements modules then partitioning these modules to lower and lower level modules until they become programmable entities. This collection of interconnected modules becomes a modular viewpoint.

3.2 Architectural styles.

An *architectural style* (a.k.a. a *pattern*) is "a set of constraints on an architecture that defines a set or family of architectures that satisfies them" [Bass, Clements, and Kazman 2003]. An architectural style can thus be seen as a meta-model, which can provide software's high-level organization (its macro-architecture) [SWEBOK 2004]. The grouping of the various architectural styles varies with the numerous sources of information about architectural styles—usually a software textbook. A popular architectural stylebook is Garlan & Shaw [1994].

(1) **General structures.** *See layers, pipes, filters, and blackboard architectures.*

    a.   **Layered architecture** — A *layered architecture* organizes the system into layers with related functionality associated with each layer. A layer provides services to the layer above it, so the lowest-level layers represent core services that are likely to be used throughout the system. An example of a layered model is a system for sharing copyright documents held in different libraries.

    b.   **Pipe-and-filters (stovepipe) architecture** — In *pipe-and-filter* style, each component has a set of inputs and a set of outputs. A component reads streams of data on its inputs and produces streams of data on its outputs. This is usually accomplished by applying a logical transformation to the input streams and computing incrementally, so that output begins before input can be completed.

    c.   **Blackboard architecture** — A *blackboard* is a place where information is posted. It only knows about itself, not about its observers. Each observer watches the blackboard via polling. When the blackboard is updated, each observer decides for itself if it needs to respond to the update. Observers do not know about each other; they only know about the blackboard.

(2) **Distributed systems.** *See client-server (two-tier), three-tier system, and broker architectures.*

    a.   **Client-server architecture** — A *client-server system* is also called a "two-tier system" or a "layered system." A layered system is organized hierarchically, each layer providing service to the layer above it and serving as a client to the layer below. In some layered systems, inner layers are hidden from all except the adjacent layer, except for certain functions carefully selected for export (i.e., when system components implement a *virtual machine* in the hierarchy).

    b.   **Three-tier architecture** — A *three-tier architecture* is also an example of a layered system—in this case, three layers. The first layer houses the HCI. The second/middle layer houses the knowledge and rules of the system, and the third layer houses any specialty systems like a database.

c. **Broker architecture** — The Common Object Request Broker Architecture (CORBA) is a *specification* of a standard architecture for object request brokers (ORBs). A standard architecture allows vendors to develop ORB products that support application portability and interoperability across different programming languages, hardware platforms, operating systems, and ORB implementations.

(3) **Interactive systems.** *An interactive system* is designed with model-view–controller or presentation-abstract-controller architecture.

a. **Model-view-controller (MVC) architectures** — The *MVC architecture* divides a given software application into three interconnected parts in order to separate internal representations from the ways that information is presented to or accepted from the user. The notions of separation and independence are fundamental to architectural design because they allow changes to be localized. The *MVC pattern* separates elements of a system, allowing them to change independently.

b. **Presentation-abstract-controller (PAC) architectures** — The *PAC architecture* is another software architectural pattern. The PAC is somewhat similar to MVC in that it separates an interactive system into three types of components responsible for specific aspects of the application's functionality.

(4) **Adaptable (interactive) systems.** Two examples of *adaptive systems* are *microkernel architecture* and *reflection architecture*.

a. **Microkernel architecture.** — A *microkernel* (also known as a μ-kernel or Samuel kernel) is the near-minimum amount of software that can provide the mechanisms needed to implement an operating system. These mechanisms include low-level address space management, thread management, and interprocess communication (IPC). If the hardware provides multiple rings or CPU modes, the microkernel is the only software executing at the most privileged level (generally referred to as supervisor or kernel mode).

b. **Reflection architecture** — The *reflection architectural* pattern provides a mechanism for changing the structure and behavior of software systems dynamically. It supports the modification of fundamental aspects, such as type structures and function call mechanisms. In this pattern, an application is split into two parts.

(5) **Other systems.** *See batch, interpreters, process control, and rule-based architectures.*

a. **Batch architecture** — The 1950s were marked by the development of rudimentary operating systems designed to smooth the transitions between jobs (a job is any program or part of a program that is to be processed as a unit by a computer). This was the start of *batch processing*, in which programs to be executed were grouped into batches.

b. **Interpreter architecture** — *Interpreters* are commonly used to build virtual machines that close the gap between the computing engine expected by the semantics of the program and the computing engine available in hardware—for example, compilers could use an interpreter to generate code [https://en.wikipedia.org/wiki/Interpreter_(computing)].

c. **Process control architecture** — This system organization is not widely recognized; nevertheless, it seems to appear within other designs. *Control-loop design* is characterized by both the kinds of components involved and the special relations that must exist among them.

   Continuous processes of many kinds convert input materials to products with specific properties by performing operations on the inputs and on intermediate products. The purpose of a control system is to maintain specific properties of the outputs of the process at given reference values (set points).

d. **Rule-based architecture** — *Rule-based systems* automate problem-solving knowledge, provide a means for capturing and refining human expertise, and prove to be commercially viable. Example: systems used in artificial intelligence [https://en.wikipedia.org/wiki/Rule-based_system].

3.3 Architectural design decisions.

Architectural design is a creative process. During the design process, SW designers make a number of fundamental decisions that profoundly affect the software and development processes. It is useful to think of the architectural design process from a decision-making perspective rather than from an activity perspective. Often, the impact on quality attributes and tradeoffs among competing quality attributes are the basis for design decisions.

Sommerville [2011] has identified a number of decisions that must be made during the design process by the system architects. Architectures must consider the following fundamental questions about the system:

(1) Is there a generic application architecture that can act as a template for the system that is being designed?

(2) How will the system be distributed across a number of cores or processors?

(3) What architectural patterns or styles might be used?

(4) What fundamental approach will be used to structure the system?

(5) How will the structural components in the system be decomposed into subcomponents?

(6) What strategy will be used to control the operation of the components in the system?

(7) What architectural organization is best for delivering the nonfunctional requirements of the system?

(8) How will the architectural design be evaluated?

(9) How should the architecture of the system be documented?

3.4 Design patterns.

Succinctly described, a *pattern* is "a common solution to a common problem in a given context" [Booch, Rumbaugh, & Jacobson 1999]. While architectural styles can be viewed as patterns describing the high-level organization of software (its macroarchitecture), other design patterns can be used to describe details at a lower, local level (its microarchitecture). Design patterns can be *creational patterns, structural patterns,* or *behavioral patterns* [SWEBOK 2004].

(1) ***Creational design patterns*** —*Creational design patterns address object-creation mechanisms* trying to create objects in a manner suitable to the situation. The form of object creation could result in design problems or added complexity to the design. Creational design patterns solve this problem by controlling this object creation.

Creational design patterns are composed of two dominant ideas: (a) encapsulating knowledge about which concrete classes the system uses and (b) hiding how instances of these concrete classes are created and combined. Examples of creational design patterns include the following [https://en.wikipedia.org/wiki/Creational_pattern]:

a. ***Abstract factory pattern*** — This pattern provides an interface for creating related or dependent objects without having to specify the object classes.

b. **Builder pattern** — This pattern separates the construction of a complex object from its representation so that the same construction process can create different representations.

c. **Factory method pattern** — This pattern allows a class to defer instantiation to subclasses.

d. **Prototype pattern** — This pattern specifies the kind of object to create using a prototypical instance and then creates new objects by cloning this prototype.

e. **Singleton pattern** — This pattern ensures that a class only has one instance and provides a global point of access to it.

(2) **Structural design patterns** — The *structural design patterns* can ease the design by identifying a simple way to realize relationships between entities. Examples of structural design patterns include the following [https://en.wikipedia.org/wiki/Structural_pattern]:

a. **Adapter pattern** — Adapts one interface for a class into one that a client expects.

b. **Bridge pattern** — Decouples an abstraction from its implementation so that the two can vary independently.

c. **Composite pattern** — A tree structure of objects where every object has the same interface.

d. **Decorator pattern** — Adds additional functionality to a class at run-time where sub classing would result in an exponential rise of new classes.

e. **Façade pattern** — Creates a simplified interface of an existing interface to ease usage for common tasks.

f. **Flyweight pattern** — Supports a high quantity of objects that share common properties to save space.

g. **Proxy pattern** — A class functioning as an interface to another thing.

(3) **Behavioral design patterns** — Behavioral *design patterns* identify common communication patterns between objects and realize these patterns. These patterns then increase flexibility in carrying out this communication. Examples of behavioral design patterns include the following [https://en.wikipedia.org/wiki/Behavioral_pattern]:

a. **Command pattern** — Encapsulates an action and its parameters.

b. **Interpreter pattern** — Implements a specialized computer language for solving a specific set of problems quickly.

c. *Mediator pattern* — Provides a unified interface to a set of interfaces in a subsystem.

d. *Memento pattern* — Provides the ability to restore an object to its previous state (rollback).

e. *Observer pattern* — Maintains a list of dependents, called observers, and notifies them automatically of any state changes.

f. *State pattern* — A clean way for an object to partially change its type at run-time.

g. *Strategy pattern* — Algorithms can be selected on the fly.

h. *Template-method pattern* — Describes the program skeleton of a program.

i. *Visitor pattern* — A way to separate an algorithm from an object.

See [http://en.Wikipedia.org.wiki/Software_de sign_pattern] for more information about design patterns.

3.5 Families of programs and frameworks.

One possible approach to allow the reuse of SEDs and components is to design *families of software* systems—a.k.a. *software product lines.*

Software product lines, or software product-line development, refers to software engineering methods, tools, and techniques for creating a collection of similar software systems from a shared set of software assets (components) using a common means of production. Carnegie Mellon Software Engineering Institute (SEI) defines a software product line as "a set of software-intensive systems that share a common, managed set of features satisfying the specific needs of a particular market segment or mission and that are developed from a common set of core assets in a prescribed way" [www.sei.cmu.edu/productlines/].

This is done by identifying the reusable and customizable components to show differences among family members. Some of the components that can be considered for use in product-line family members are [SWEBOK 2004]:

(1) Hardware components.

(2) Research and development (R&D).

(3) User training books.

(4) Software operating systems.

(5) Software applications.

(6) Manufacturing techniques.

(7) Manufacturing methods.

(8) Maintenance techniques.

(9) Maintenance manuals.

(10) User manuals.

In *object-oriented programming*, a key related notion is that of the *framework*: a partially complete software subsystem that can be extended by appropriately instantiating specific plug-ins (a.k.a. "hot spots").

Recent advances in the software product-line field have demonstrated that narrow and strategic application of these concepts can yield order of magnitude improvements in software engineering capability.

4. Detailed Design

4.1 Introduction.

A software detailed design is a representation or model of the software system to be created. The model should provide the precise design information needed for planning, analysis, and implementation of the software system. It should represent a partitioning of the system into design entities and describe the important properties and relationships among those entities.

The design description model used to represent a software system can be expressed as a collection of design entities, each possessing properties and relationships. To simplify the model, the properties and relationships of each design entity are described by a standard set of attributes. The design information needs of project members are satisfied through identification of the entities and their associated attributes. A design description is complete when the attributes have been specified for all the entities [IEEE Std. 1016-1998].

4.2 Design entities.

A *design entity* is an element (component) of a design that is structurally and functionally distinct from other elements and that is separately named and referenced. Design entities result from a decomposition of the software system requirements. The objective is to divide the system into separate components that can be considered, implemented, changed, and tested with minimal effect on other entities [IEEE Std. 1016-1998].

The number and type of entities required to partition a design are dependent on a number of factors, such as the complexity of the system, the design technique used, and the programming environment.

Although entities are different in nature, they possess common characteristics. Each design entity will have a name, purpose, and function. There are common relationships among entities such as interfaces or shared data. The common characteristics of entities are described by design entity attributes.

4.3 Design entity attributes.

A *design entity attribute* is a named characteristic or property of a design entity. It provides a statement of fact about the entity.

Design entity attributes can be thought of as questions about design entities. The answers to those questions are the values of the attributes. All the questions can be answered, but the content of the answer will depend upon the nature of the entity. The collection of answers provides a complete description of an entity.

The list of design entity attributes presented in this subclause is the minimum set required for all SDDs.

All attributes shall be specified for each entity. Attribute descriptions should include references and design considerations such as tradeoffs and assumptions when appropriate. In some cases, attribute descriptions may have the value *none.* When additional attributes are identified for a specific software project, they should be included in the design description. The attributes and associated information items are defined in Paragraphs 4.3.1 through 4.3.8.

4.3.1 Identification. *The name of the entity.* Two entities shall not have the same name. The names for the entities may be selected to characterize their nature. This will simplify referencing and tracking in addition to providing identification specification.

4.3.2 Function. *A statement of what the entity does.* The function attribute shall state the transformation applied by the entity to inputs to produce the desired output. In the case of a data entity, this attribute shall state the type of information stored or transmitted by the entity.

4.3.3 Subordinates. *The identification of all entities composing this entity.* The subordinates attribute shall identify the *composed of* relationship for an entity. This information is used to trace requirements to design entities and to identify parent/child structural relationships through a software system decomposition.

4.3.4 Dependencies. *A description of the relationships of this entity with other entities.* The dependencies attribute shall identify the *uses* or *requires the presence of* relationship for an entity. These relationships are often graphically depicted by structure charts, data flow diagrams, and transaction diagrams.

This attribute shall describe the nature of each interaction including such characteristics as timing and conditions for interaction. The interactions may involve the initiation, order of execution, data sharing, creation, duplicating, usage, storage, or destruction of entities.

4.3.5 Interface. *A description of how other entities interact with this entity.* The interface attribute shall describe the *methods* of interaction and the *rules* governing those interactions. The methods of interaction include the mechanisms for invoking or interrupting the entity, for communicating through parameters, common data areas or messages, and for direct access to internal data. The rules governing the interaction include the communications protocol, data format, acceptable values, and the meaning of each value.

This attribute shall provide a description of the input ranges, the meaning of inputs and outputs, the type and format of each input or output, and output error codes. For information systems, it should include inputs, screen formats, and a complete description of the interactive language.

4.3.6 Resources. *A description of the elements used by the entity that are external to the design.* The resources attribute shall identify and describe all of the resources *external* to the design that are needed by this entity to perform its function. The interaction rules and methods for using the resource shall be specified by this attribute.

This attribute provides information about items such as physical devices (printers, disc-partitions, memory banks), software services (math libraries, operating system services), and processing resources (CPU cycles, memory allocation, buffers).

The resources attribute shall describe usage characteristics such as the process time at which resources are to be acquired, and sizing to include quantity and physical sizes of buffer usage. It should also include the identification of potential race and deadlock conditions as well as resource management facilities.

4.3.7 Processing. *A description of the rules used by the entity to achieve its function.* The processing attribute shall describe the algorithm used by the entity to perform a specific task and shall include contingencies. This description is a refinement of the function attribute. It is the most detailed level of refinement for this entity.

This description should include timing, sequencing of events or processes, prerequisites for process initiation, priority of events, processing level, actual process steps, path conditions, and loop back or loop termination criteria. The handling of contingencies should describe the action to be taken in the case of overflow conditions or in the case of a validation check failure.

4.3.8 Data. *A description of data elements internal to the entity.* The data attribute shall describe the method of representation, initial values, use, semantics, format, and acceptable values of internal data.

The description of data may be in the form of a data dictionary that describes the content, structure, and use of all data elements. Data information shall describe everything pertaining to the use of data or internal data structures by this entity. It shall include data specifications such as formats, number of elements, and initial values. It shall also include the structures to be used for representing data such as file structures, arrays, stacks, queues, and memory partitions.

The meaning and use of data elements shall be specified. This description includes such things as static versus dynamic, whether it is to be shared by transactions, used as a control parameter, or used as a value, loop iteration count, pointer, or link field. In addition, data information shall include a description of data validation needed for the process.

5. User Interface Design

Computer system design encompasses a spectrum of activities, from hardware design to user interface design. While specialists are often employed for hardware design and for the graphic design of web pages, only large organizations normally employ specialist interface designers for their application software. Software engineers must often take responsibility for user interface design as well as for the design of the software to implement that interface [Sommerville 2007]. *Note: SWEBOK uses a different outline for user interface design.*

5.1 General HCI design principles.

The Association for Computing Machinery (ACM) defines HCI as "a discipline concerned with the design, evaluation, and implementation of interactive computing systems for human use and with the study of major phenomena surrounding them" [ACM SIGCHI 1996].

Human–computer interface (HCI) is the study of interactions between people (users) and computers. These interactions can be interactions between users and computers that occur at the user interface. They can include both software and hardware. For example, characters or objects displayed by software on a personal computer's monitor, input received from users via hardware peripherals such as keyboards and mice, and other user interactions with large-scale computerized systems such as aircraft and trains [http://en.wikipedia.org /wiki/Human_computer_interaction].

HCI design principles vary greatly, depending upon who is reporting. The following list of design principles by Sommerville appears to be the best, the most understandable, and among the shortest [Sommerville 2007]:

(1) **User familiarity** — The interface should use terms and concepts drawn from the experience of those who will make the most use of the system.

(2) **Learnability** — The software should be easy to learn so that the user can rapidly start working with the software.

(3) **Consistency** — The interface should be consistent in that, wherever possible, comparable operations should be activated the same way.

(4) **Minimal surprise** — Users should never be surprised by the behavior of a system.

(5) **Recoverability** — The interface should include mechanisms to allow users to recover from errors.

(6) **User guidance** — The interface should provide meaningful feedback when errors occur and provide context-sensitive user help.

(7) **User diversity** — The interface should provide appropriate interaction facilities for different types of system users.

5.2 Use of modes.

A *mode* is defined as a particular form or variation of something. An HCI mode is a distinct method of operation within a computer program in which the same input can produce different perceived results, depending on the mode of the computer program. For example, "caps lock" sets an input mode in which typed letters are uppercase by default; the same typing produces lowercase letters when not in the caps lock mode. Heavy use of modes often reduces the usability of a user interface, as the user must expend effort to remember current mode states and switch between mode states as necessary [https://en.wikipedia.org /wiki/User_interface_design].

5.3 Use of navigation.

Navigation is the ability to move efficiently and effectively through a document, such as moving through web pages that are linked together to find a particular word or paragraph. There are several standard navigational layouts—*linear, hierarchical* and *webbed* [http://en.wikipedia.org/wiki/Interface_(computing)].

(1) **Linear navigation systems** — A *linear navigation system* allows users to navigate through the interface in one direction. Linear systems are normally non or minimally interactive, with Start, Next, and Back buttons but no others. Linear systems are the most structured of all the navigation categories.

(2) **Hierarchical based site** — A *hierarchical based site* is similar to a family

tree in that every page has a parent page. Each page has only one page leading to it. The structure of sites of this type is very rigid, but it can be useful for organizations with discrete departments requiring one section each. *Breadcrumb trails* can also be used in a hierarchical site.

(3) **Webbed topology** — A *webbed topology* allows the user to navigate in a more fluid fashion. The range of webbed topologies is quite broad, ranging from almost hierarchical with additional links to those in which each page will link to others of similar or related topics, but without an underlying categorization by topic.

5.4 Coding techniques and visual design.

Coding and visual design include such techniques as the use of different colors, icons, and fonts.

(1) **Color** — *Color* can improve user interfaces by helping users to understand and manage complexity. However, it is easy to misuse color and to create user interfaces that are visually unattractive and error-prone. Schneiderman [1998] developed 14 key guidelines for the effective use of color in user interfaces. The six most important of these are as follows [Sommerville 2007]:

a. **Limit the number of colors employed and be conservative as to how they are used** — You should not use more than four or five separate colors in a window and no more than seven in a system interface. If you use too many, or if they are too bright, the display may be confusing. Some users may find masses of color disturbing and visually tiring. User confusion is also possible if colors are used inconsistently.

b. **Use color change to show a change in system status** — If a display changes color, this should mean that a significant event has occurred. Thus, in a fuel gauge, you could use a change of color to indicate that fuel is running low. Color highlighting is particularly important in complex displays with hundreds of distinct entities being displayed.

c. **Use color-coding to support the task users are trying to perform** — If they have to identify anomalous instances, highlight these instances; if similarities are also to be discovered, highlight these using a different color.

d. **Use color-coding in a thoughtful and consistent way** — For instance, if one part of a system displays error messages in red, all other parts should do likewise. Red should not be used for anything else. If it is, the user may interpret the red display as an error message.

e. ***Be careful about color pairings*** — Because of the physiology of the eye, people cannot focus on red and blue simultaneously. Eyestrain is a likely consequence of a red-on-blue display. Other color combinations may also be visually disturbing or difficult to read.

f. ***In general, you should use color for highlighting, but you should not associate meanings with particular color*** — About 10% of men are color-blind and may misinterpret the meaning. Human color perceptions are different, and there are different conventions in different professions about the meaning of particular colors. Users with different backgrounds may unconsciously interpret the same color in different ways. For example, to a driver, red usually means danger or stop. But, to a chemist, red means hot. So do not use color for meaning.

(2) ***Icons*** — *Icons* are pictographic representations of data or processes within a computer system, which have been used to replace commands and menus as the means by which the computer supports the end user. Icons have been applied principally to graphics-based interfaces on operating systems, networks, and document-processing software.

(3) ***Fonts*** — A study of *fonts* was conducted at Wichita State University [Bernard, Liao & Mills 2001] to determine the impact of font size and style on legibility, reading time and general preference when read by an older population. The study involved volunteers reading text passages containing two serif and sans serif fonts at 12- and 14-points.

Two types of fonts were used—the serif fonts Georgia and Times New Roman, and the sans serif fonts Arial and Verdana. Both the Times New Roman and Arial fonts were originally developed for printing. These fonts are the most common fonts of their respective font type used today. The Georgia and Verdana, however, were developed specifically for optimized viewing on a computer screen.

Several observations can be made from these findings:

a. First, 14-point fonts were found to be more legible and promote faster reading and were preferred over 12-point fonts.

b. Second, at the 14-point size, serif fonts tended to support faster reading. (Serif fonts, however, were generally not preferred over the sans serif fonts.)

c. Third, there was essentially no difference in readability between the computer fonts and the print fonts.

d. Therefore, 14-point fonts are recommended for presenting online text to older readers. However, a compromise must be made in deciding which font type to use. If speed of reading is paramount, then serif fonts are recommended. However, if visual preference is important, then sans serif fonts are recommended.

5.5 Response time and feedback.

Slow response times and *difficult navigation* are the most common complaints of Internet users. After waiting past a certain "attention threshold," users bail out to look for a faster site. Of course, exactly where that threshold is depends on many factors. For example, how compelling is the experience? Is there effective feedback?

The following is a set of response and feedback times proposed by Dr. Ben Shneideman https://www.cs.umd.edu/~ben/papers/Shneiderman1984Response.pdf]

(1) Load in under 8.6 seconds (non-incremental display).

(2) Decrease these load times by 0.5 to 1.5 seconds for dynamic transactions.

(3) Minimize the number of steps needed to accomplish tasks to avoid cumulative frustration from exceeding user-time budgets.

(4) Load in under 20 to 30 seconds (incremental display), with useful content within 2 seconds.

(5) Provide performance information.

(6) Equalize page download times to minimize delay variability.

5.6 Design modalities.

Multimodal interfaces support perceptual capabilities (e.g., auditory, speech, and visuals) as a means of facilitating human interaction with computers [Sears & Jacko 2007]. Examples are *menu-driven, command-line driven*, and *event-driven interfaces.* Other interface support tools that make interacting with computer systems easier include *forms* and *question-answer interfaces.*

(1) **Menu-driven interface** — The *menu-driven interface* consists of a series of screens that are navigated by choosing options from lists (i.e., menus). (Here, "menu" does not refer to pull-down menus, but to lists of options on the screen that lead to other screens.) Because of their simplicity, menu-driven interfaces are commonly used for walk-up-and-use systems, such as information kiosks and ATMs. Websites are also often designed

with the same basic navigation principle, where navigation bars substitute for menus.

(2) **Command-line interface** — The *command-line interface* is a means of operating a computer by typing a text command in response to an on-screen prompt and hitting the Enter or Return key to issue the command. The computer then processes the command, displays whatever output is appropriate, and presents another prompt for the next command. Typical commands are to run a program, enter a text editor, list files, and change directories. This mode of interaction is common (e.g., in the traditional DOS and UNIX operating systems).

(3) **Event-driven interface** — An *event-driven interface* is common to most modern operating systems where the user can initiate actions at any time; the system responds to user "events," such as typing, mouse movements, or mouse clicks.

(4) **Forms dialogue boxe interface** — A user employs a *forms dialogue box interface* to communicate with the system by filling in an on-screen form (e.g., a data entry form on a database). Design of the form must be clearly worded and presented, and color and highlights can be used. Form filling enables experienced users to enter data quickly and is user-friendly to the less experienced user.

(5) **Question-answer interface** — In a *question-answer interface*, the application asks questions, and when the user provides answers containing all necessary data, the application gives the results. This interface is sometimes called "walkthrough and use" or "interview" applications.

5.7 Human-computer interface (HCI) design methods.

Schneiderman's "Eight Golden Rules of Interface Design" were obtained from the text *Designing the User*. Schneiderman proposed this collection of principles that are derived heuristically from experience and are applicable in most interactive systems after being properly refined, extended, and interpreted [http://faculty.washington.edu/jtenenbg/courses/360/f04/sessions/Schneiderman.GoldenRules.html].

(1) **Strive for consistency** — *Consistent sequences of actions* should be required in similar situations; identical terminology should be used in prompts, menus and help screens; consistent commands should be employed throughout.

(2) **Enable frequent users to use shortcuts** — As the *frequency of use* increases, so do the user's desires to reduce the number of interactions

and to increase the pace of interaction. Abbreviated function keys, hidden commands, and macro-facilities can be very helpful to a user.

(3) ***Offer informative feedback*** — For every *user action*, there should be some system feedback. For frequent and minor actions, the response can be modest, while for infrequent and major actions, the response should be more substantial.

(4) ***Design dialog to yield closure*** — *Sequences of actions* should be organized into groups with a beginning, middle, and end. The informative feedback at the completion of a group of actions gives operators the satisfaction of accomplishment, a sense of relief, the signal to drop contingency plans and options from their minds, and an indication that the way is clear to prepare for the next group of actions.

(5) ***Offer simple error handling*** — As much as possible, design the system so the user cannot make a *serious error*. If an error is made, the system should be able to detect the error and offer simple, comprehensible mechanisms for handling the error.

(6) ***Permit easy reversal of actions*** — Allowing the user to *reverse* an action relieves anxiety, since the user knows that errors can be undone; it thus encourages exploration of unfamiliar options. The units of reversibility may be a single action, a data entry, or a complete group of actions.

(7) ***Support internal locus of control*** — Experienced operators strongly desire the sense that *they are in charge of* the system and that the system responds to their actions. Design the system to make users the initiators of actions rather than the responders.

(8) ***Reduce short-term memory load*** — The limitation of human information processing in *short-term memory* requires that displays be kept simple, multiple page displays be consolidated, window motion frequency be reduced, and sufficient training time be allotted for codes, mnemonics, and sequences of actions.

5.8 Multimedia.

Multimedia is media and content that uses a combination of different content forms. The term can be used as a noun meaning a medium with multiple content forms or as an adjective describing a medium as having multiple content forms. The term is used in contrast to "media," referring to only traditional forms of printed or hand-produced material. Multimedia includes a combination of text, audio, still images, animation, video, and interactivity content forms.

Multimedia is usually recorded and played, displayed, or accessed by information content processing devices such as computerized and electronic devices, but it can also be part of a live performance. Multimedia (as an adjective) also describes electronic media devices used to store and experience multimedia content [http://en.wikipedia.org/wiki/Multimedia].

(1) **Input and output** — *Input/output* (written I/O), refers to the communication between an information processing system (such as a computer) and the outside world—possibly a human, or another information processing system. Inputs are the signals or data received by the system, and outputs are the signals or data sent from it.

The term can also be used as part of an action: to "perform I/O," or to perform an input or output operation. I/O devices are used by a person (or other system) to communicate with a computer. For instance, a keyboard or a mouse may be an input device for a computer, while monitors and printers are considered output devices for a computer. Devices for communication between computers, such as modems and network cards, typically serve for inputs and outputs.

Devices must be constructed for mediating between humans and machines. Examples are:

a. **Input devices** — *Input devices* are mechanics of particular devices, performance characteristics (human and system), devices for the disabled, handwriting and gestures, speech input, eye tracking, and exotic devices (e.g., electro-encephalo-gram (EEG) and other biological signals).

b. **Output devices** — *Output devices* are mechanics of particular devices, vector and raster devices, frame buffers and image stores, canvases, event handling, performance characteristics, devices for the disabled, sound and speech output, 3D displays, motion devices (e.g., flight simulators), and exotic devices.

c. **Characteristics of I/O devices** — *I/O devices* are used to accept or provide data into a computing system; for example, weight, portability, bandwidth, and sensory modality.

Note that the designation of a device as either input or output depends on the perspective. Mice and keyboards take as input physical movement that the human user outputs and converts this movement into signals that a computer can understand. Hence, the output from these devices is input for the computer. Similarly, printers and monitors take input signals that are output by a computer.

The I/O devices then convert these signals into representations that human users can see or read. For a human user, the process of reading or seeing these representations comprises the reception of input [http://en. wikipedia.org/wiki/Input/output].

(2) **Natural languages** — *Natural language*s deal with computer systems that interpret the languages used by humans. The ultimate goal is to be able to communicate with your computer as you would with another person. Unfortunately, natural language, which is the easiest for humans to learn, is the hardest for computers to learn.

The Natural Language Processing Group based in Redmond, Washington, and described at [http://research.microsoft.com/enus/groups/nlp/] is working to develop algorithms and statistical models that can efficiently interpret natural language.

The group's advancements have been integrated into applications including information recovery, text analyzing, question answering, gaming, and many others. As the group's work progresses, they anticipate it will enable people to communicate with computers through natural language [https://wikispaces.psu .edu/display/331Grp1/Natural+language+HCI +Information].

(3) **Sound** — *Sound* recording and reproduction is an electrical or mechanical inscription and recreation of sound waves, such as spoken voice, singing, instrumental music, or sound effects. The two main classes of sound recording technology are *analog recording* and *digital recording*.

   a. **Acoustic analog recording** — *Acoustic analog recording* is achieved by a small microphone diaphragm that can detect changes in atmospheric pressure (acoustic sound waves) and record them as a graphic representation of the sound waves on a medium such as a phonograph.

   b. **Digital recording and reproduction** — *Digital recording and reproduction* converts the analog sound signal picked up by the microphone to a digital form by a process of digitization, allowing it to be stored and transmitted by a wider variety of media [http://en.wikipedia.org/wiki/Sound_recording_and_reproduction].

(4) **Text** — *Text* (writing) is the representation of language in a textual medium with a set of signs or symbols (known as a writing system). It is distinguished from illustration, such as cave drawing and painting, and from non-symbolic preservation of language via nontextual media, such as a magnetic tape [http://en.wikipedia.org/wiki/Writing].

(5) **Voice** — *Voice* (or vocalization) is the sound produced by humans and other vertebrates using the lungs and the vocal folds in the larynx or voice box. Voice is not always produced as speech, however. Infants

babble and coo; animals bark, moo, whinny, and meow; and humans laugh, sing, and cry.

Voice is generated by airflow from the lungs as the vocal folds are brought close together. When air is pushed past the vocal folds with sufficient pressure, the vocal folds vibrate. If the vocal folds in the larynx did not vibrate normally, speech could only be produced as a whisper. Your voice is as unique as your fingerprint. It helps define your personality, mood, and health [http://www.nidcd.nih.gov/health/voice/Pages/whatis_vsl.aspx].

Voice is a major means of multimedia. Voice can be applied to many other multimedia types to increase understanding by the receiver (listener).

(6) **Web page** — Each *web page* (also spelled webpage) represents various types of information presented to the visitor in an aesthetic and readable manner. Most web pages are available on the World Wide Web, which makes them widely accessible to the Internet public. The information on a web page is displayed online with the help of a web browser, which connects with the server where the website's contents are hosted through the Hypertext Transfer Protocol (HTTP) [http://www.ntchosting.com/internet/web page.html].

5.9 Metaphors and conceptual models.

Metaphors are devices that express an abstract concept through analogy. The use of metaphors allows unfamiliar and abstract concepts to be more readily grasped and understood.

An example of a metaphor in everyday speech is talking about time as if it is money or currency. Time is an abstract concept, and by using a metaphor to make it more familiar and understandable, we can talk about it more freely. By using this money metaphor in relation to time, it has become normal for us to save, spend, give, waste, and borrow time.

Metaphors are important within HCI because they allow users to apply their understanding of everyday objects and situations to help them understand concepts within a computing environment.

The desktop metaphor is one that has been used from an early stage by the Mac Windows System, which then of course led to the Windows Operating System by Microsoft. It is important to note that not all of the functionality of a real-world desktop can be transformed into its virtual counterpart. Novice users who are accustomed to a certain behavior in their real-world desktops can be surprised when things are not quite the same with their computer desktops. Such HCI metaphors, like all metaphors, should not be taken literally.

A *conceptual model* describes the way a system is designed to be understood. A *mental model* is the way the user actually understands the system. A good conceptual model that is applied properly in the design of a system will enable a user to develop a good mental model to associate with the system.

There are metaphors and mental models that can be used in Instructional System Design (ISD) to help users gain a good understanding of the system. The user's mental model of a system is developed by viewing and/or experiencing the system, including its visible functionality and structure.

For an interactive systems designer, it is good practice to start with a desired mental model and then develop the interface with the intention of conveying that mental model explicitly to the user through a conceptual model. Parts of the system that may clash with the conceptual model can be hidden from the user in order to maintain a good conceptual model, which leads to ease of use [http://www.computingstudents.com/notes/interactive_systems/metaphors_c onceptula_models.php].

5.10 Psychology of HCI.

Cognitive psychologists who work in the software industry typically find themselves designing and evaluating complex software systems to aid humans in a wide range of problem domains, like word processing, interpersonal communications, information access, finance, remote meeting support, air traffic control, or even gaming situations.

In these domains, the technologies and the users' tasks are in a constant state of flux, evolution, and co-evolution. Cognitive psychologists working in HCI design may try to start from first principles in developing these systems, but they often encounter novel usage scenarios for which no guidance is available. For this reason, we believe that there is not as much application of theories, models, and specific findings from basic psychological research to user interface (UI) design, as one would hope. However, several analysis techniques and some guidelines generated from the literature are useful [Dumais and Czerwinski 2001].

6. Software Design Notations

Many notations and languages exist to represent SED artifacts. Some are used mainly to describe a design's *structural organization* and others to represent *software behavior*. Certain notations are used mostly during architectural design and others mainly during detailed design, although some notations can be used in both steps. In addition, some notations are used mostly in the context of specific methods. Here, they are categorized into notations for describing the structural (static) view and notations for describing the behavioral (dynamic) view [SWEBOK 2004].

6.1 Structural descriptions (static view).

The following notations are mostly (but not always) graphical, and they describe and represent the structural aspects of an SED; that is, they describe the major components and how they are interconnected (static view) [SWEBOK 2004]:

(1) **Architecture description languages (ADLs)** — *Textual*, often formal languages used to describe software architecture in terms of components and connectors.

(2) **Class and object diagrams** — *Class and object diagrams* are used to represent a set of classes (and objects) and their interrelationships.

(3) **Component diagrams** — *Component diagrams* are used to represent a set of components ("physical and replaceable parts of a system that conform to and provide the realization of a set of interfaces") [Booch, Rumbaugh, & Jacobson 1999] and their interrelationships.

(4) **Class responsibility collaborator (CRC) models** — *CRC models* are used to denote the names of classes, their responsibilities and collaborating components' names [http://www.agilemodeling.com/artifacts /crcModel.htm].

(5) **Deployment diagrams** — *Deployment diagrams* are used to represent a set of physical nodes and their interrelationships, and thus to model the physical aspects of a system.

(6) **Entity-relationship diagrams (ERDs)** — *ERDs* are used to represent conceptual models of data stored in information systems.

(7) **Interface description languages (IDLs)** — *Programming-like languages* are used to define the interfaces (names and types of exported operations) of software components.

(8) **Structure charts** — *Structure charts* are used to describe the calling structure of programs (which module calls, and is called by, which other modules). A calling structure is the activation or initiating of a subroutine or function by another subroutine.

6.2 Behavioral descriptions (dynamic view).

The following notations and languages, some graphical and some textual, are used to describe the dynamic behavior of software and components. Many of these notations are useful mostly, but not exclusively, during detailed design [SWEBOK 2004]:

(1) **Activity diagrams** — *Activity diagrams* are used to show the control flow between activities.

(2) **Collaboration diagrams** — *Collaboration diagrams* are used to show the interactions that occur among a group of objects, where the emphasis is on the objects, their links, and the messages exchanged on these links.

(3) **Data flow diagrams (DFDs)** — *DFDs* are used to show data flow among a set of processes.

(4) **Decision tables and diagrams** — *Decision tables* and *diagrams* are used to represent complex combinations of conditions and actions.

(5) **Flowcharts and structured flowcharts** — *Flow charts* are used to represent the flow of control and the associated actions to be performed.

(6) **Sequence diagrams** — *Sequence diagrams* are used to show the interactions among a group of objects, with emphasis on the time of messages.

(7) **State transition and state-chart diagrams** — *State transition diagrams* are used to show the control flow from state to state in a state machine.

(8) **Formal specification languages** — *Textual languages* that use basic notions from mathematics (e.g., logic, set, and sequence) to rigorously and abstractly define software component interfaces and behavior, often in terms of pre- and post-conditions.

(9) **Pseudocode and program design languages (PDLs)** — *Structured programming-like languages* used to describe, generally at the detailed design stage, the behavior of a procedure or method.

7. Software Design Strategies and Methods

Various general strategies and more focused methods can help guide the design process. In contrast to general strategies, methods are more specific in that they generally suggest and provide a set of notations to be used with the method, a description of the process to be used when following the method, and a set of guidelines in using the method. Such methods are useful as a means of transferring knowledge and as common frameworks for teams of software engineers [SWEBOK 2004].

7.1 General strategies.

Some often-cited examples of general strategies useful in the design process are *divide and conquer, stepwise refinement, top-down versus bottom-up strategies, data abstraction,* and *information hiding,* use of *heuristics,* use of *patterns and pattern languages,* and an *iterative* and *incremental* approach [SWEBOK 2004].

(1) **Divide and conquer** — *Divide and conquer* (D&C) is an important algorithm design paradigm based on multi branched recursion. A D&C algo-

rithm works by recursively breaking down a problem into two or more subproblems of the same (or related) type, until these become simple enough to be solved directly. The solutions to the subproblems are combined to give a solution to the original problem.

This technique is the basis of efficient algorithms for all kinds of problems, such as sorting (e.g., quicksort and merge sort), multiplying large numbers, syntactic analysis (e.g., top-down parsers), and computing the discrete Fast Fourier Transforms (FFTs) [http://en.wikipedia.org/wiki /Divide_and_conquer_algorithm].

(2) **Stepwise refinement** — *Stepwise refinement* is a way of developing a computer program by first describing general functions, then breaking each function down into successive steps until the whole program is fully defined. Stepwise refinement is also called *"top-down design."*

(3) **Top-down versus bottom-up strategies** — A *top-down approach* (a.k.a. *stepwise design*) is essentially the breaking down of a system to gain insight into its compositional subsystems. In a top-down approach, an overview of the system is formulated, specifying but not detailing any subsystems. Each subsystem is then refined into greater detail until the entire specification is reduced to base components. This approach starts with a top-view of the system requirements and through partishing, develops a set of interlocking segments that are technically equivalent to the original requirements.

A *bottom-up approach* is the piecing together of systems to give rise to more complex systems, thus making the original systems subsystems of the emergent system. In a bottom-up approach, the individual base elements of the system are first specified in detail. These elements are then linked together to form larger subsystems, which then in turn are linked, sometimes in many levels, until a complete top-level system is formed. This strategy often resembles a "seed" model, whereby the beginnings are small but will eventually grow [http://en.wikipedia.org /wiki/Top-down_and_bottom-up_design].

(4) **Information hiding** — In 1972, David Parnas introduced the idea of *information hiding.* He defined information hiding as a way in which clients could be shielded from internal program workings. Information hiding has the capability of containing a software process in a module or subroutine and hiding the internal mechanisms from the user (sometimes called a software "black box").

The only parts of the process available to the user are the input requirements and the output results. The process of converting the input to the output is hidden from the user. This approach hides the details of a process from a user who does not need to know how the answer was derived. The approach greatly simplifies the software development process. In addition, this approach can also promote software reuse.

(5) *Heuristics* — *Heuristics* are strategies using readily accessible, though loosely applicable, information to control problem solving in human beings and machines [http://en.wikipedia.org/wiki/Heuristic]. A heuristic is an experience-based technique for problem solving, learning, and discovery that gives a solution, which is not guaranteed to be optimal.

Where the exhaustive search is impractical, heuristic methods are used to speed up the process of finding a satisfactory solution via mental shortcuts to ease the cognitive load of making a decision. Other terms used in conjunction with heuristics include *rules of thumb*, *educated guesses*, *intuitive judgment*, *stereotyping*, *common sense*, and so forth.

(6) *Patterns and pattern languages* — In general, a *pattern* is a discernible regularity in the world or in a human-made design. As such, the elements of a pattern repeat in a predictable manner. A geometric pattern is formed using geometric shapes and typically repeats, like a wallpaper. An SED pattern is a known solution to a class of programming problems.

A *pattern language* is a method of describing good design practices within a field of expertise. The term was coined by architect Christopher Alexander and popularized by his book *A Pattern Language* [Alexander et al. 1977]. Advocates of this design approach claim that ordinary individuals can solve very large, complex design problems.

Like all languages, a pattern language has a vocabulary, syntax, and grammar. This simplifies the design work because designers can start the process from any part of the problem they understand and work toward the unknown parts. At the same time, if the pattern language has worked well for many projects, there is reason to believe that the design problem can still complete the design process, and the result will be usable [http://en.wikipedia.org/wiki/Pattern].

(7) *Iterative approach* — *Iteration* in computing is the repetition of a block of statements within a computer program. It can be used both as a general term, synonymous with "repetition," and to describe a specific form of repetition with a mutable state.

Iteration can be the repetition of a mathematical or computational procedure applied to the result of a previous application, typically as a means of obtaining successively closer approximations to the solution of a problem. Each step is called an "iteration."

The calling of one function from another immediately suggests the possibility of a function calling itself. This process is called "recursion." A powerful general-purpose programming technique, recursion is the key to numerous critically important computational applications [https://en .wikipedia.org/wiki/Iteration].

(8) **Incremental approach** — The *incremental approach* is a method of software development where the model is designed, implemented, and tested incrementally until the product is finished. It involves both development and maintenance. The product is defined as finished when all of its requirements are satisfied [https://en.wikipe dia.org/wiki/Incre mental_build_model].

7.2 Function-oriented (structured) design.

This is one of the classical methods of SED, where decomposition centers on identifying the major software functions and then elaborating and refining them in a top-down manner. *Structured design* is generally used after *structured analysis*, thus producing, among other things, data flow diagrams and associated process descriptions. Researchers have proposed various strategies (for example, transformation analysis and transaction analysis) and heuristics (for example, fan-in/fan-out, and scope of effect versus scope of control) to transform a data flow diagram (DFD) into a software architecture generally represented as a structure chart [SWEBOK 2004].

(1) **Transaction analysis** — The main purpose of *transaction analysis* is to separate the components of a large design into a network of cooperating subsystems that can then be combined. This is done by identifying the transactions that are involved in the problem as a whole. The DFD components that correspond to each transaction are then grouped together and used as input to a transform analysis step. The resulting structure charts are then recombined to provide the design model for the complete system [Budgen 2003].

(2) **Transform analysis** — *Transform analysis*, the key transformation of function-oriented design, is performed on the DFD that is created to describe a given transaction. It is in this step that the designer takes the nonhierarchical model constructed to describe the problem, modelled around the flow of data between operations, and transforms this to create a description of the structure of a computer program. This is modeled in terms of the hierarchy formed dependent upon how subpro-

grams are invoked, together with the flow of information created with parameters and shared data structures [Budgen 2003].

    a.   The first action of the designer is to identify the operation or "bubble" that acts as the central transform in the DFD. [*See Paragraph 1.7.(3) to define bubble*] The central transform is the bubble that lies at the center of input and output data flow—where these are considered to have their most abstract form (given that they take their most concrete form when interacting with physical I/O devices). However, it is not always possible to identify a clear candidate to act as the central transform, and in these cases, the recommended practice is to create one by adding another bubble in the position where a central transform should occur.

    b.   A feature of this process that sometimes gives conceptual difficulty is that the flow arrows on the arcs seem to change direction when this transformation is made. This is because the arc in a DFD depicts data flow, while the arc in a structure chart depicts control flow (via subprogram invocation). The latter is added in this design step, and the former is subsumed into the data flow that is conducted via the parameters of the subprograms.

(3) **Structured analysis** —*Structured analysis* was popular in the 1970s and is still in use in 2016. *Structured analysis* consists of modeling the system data-flow diagrams (a.k.a. a bubble chart). A bubble is used to represent a requirements process. The process is represented on paper as a circle (i.e., a bubble). Each bubble can be partitioned into a set of bubbles, each one representing a subset of the original process. The flow of data can be from a bubble to a data store and then to another bubble or between bubbles.

Structured analysis is part of what is called the "*structured method*" includes *context diagrams, dataflow diagrams, mini-specs,* and *data dictionaries*; many of these techniques are credited to Edward Yourdon [1989].

(4) **A *data flow diagram* (DFD)** — A DEF is a graphical representation of the "flow" of data through an information system (*See also Chapter 2, Paragraph38.3(2).*)

(5) **Software design heuristics** — As stated earlier in this chapter [[*Paragraph 7.1(5)*], a *heuristic* refers to experience-based techniques for problem solving, learning, and discovery that provide a solution to a design problem (however, this approach is not guaranteed to be optimal).

Where the exhaustive search is impractical, heuristic methods are used to speed up the process of finding a satisfactory solution via mental shortcuts to ease the cognitive load of making a decision. Other terms include using a *rule of thumb,* an *educated guess,* an *intuitive judgment, stereotyping,* or *common sense.* The following are some SWE heuristics:

a. ***Span of control (fan-in/fan-out)*** — The *span of control* of a module is the number of its immediately superordinate (i.e., parent or boss) modules (*See Figure 5*). The designer should strive for high fan-out at the lower levels of the hierarchy and low fan-in at the higher levels of the module.

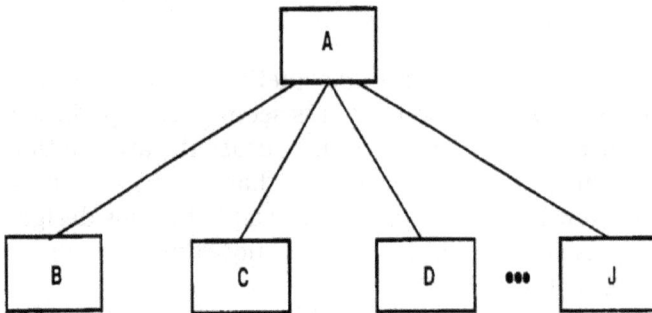

Figure 5: Illustration of span of control

The process is called "top-down design," or "functional decomposition." Programmers use a structure chart to build a program in a manner similar to an architect using a blueprint to build a house.

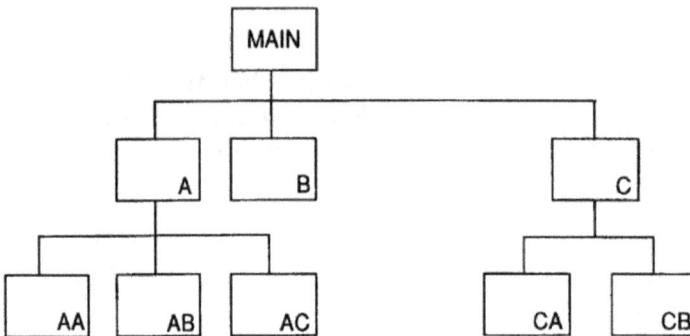

Figure 6 : Illustration of a structure chart

The chart is drawn and used for the client and the designers to communicate [http://en.wikipedia.org/wiki/Struture_chart].

This simply means that normally there are common low-level functions that should be identified and made into common modules to reduce redundant code and increase maintainability. High fan-in can also increase portability if, for example, all I/O handling is done in common modules.

The process is called "top-down design," or "functional decomposition." Programmers use a structure chart to build a program in a manner similar to an architect using a blueprint to build a house. The chart is drawn and used for the client and the designers to communicate [http://en.wikipedia.org/wiki/Struture_chart].

b. ***Scope of effect versus scope of control*** — The *scope of effect* of a module is defined as all of the modules affected by a decision made by the module The *scope of control* of a module is defined as the module itself and all of its subordinates (i.e., those modules it calls and all the modules called by it). (*See Figure 7.*)

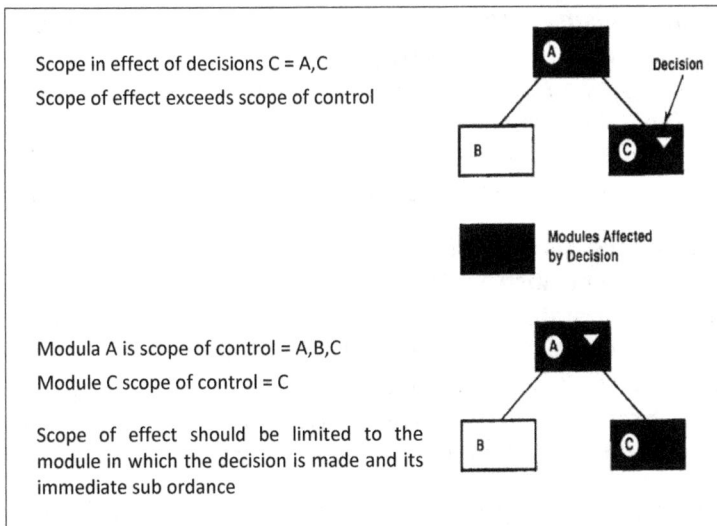

Scope in effect of decisions C = A,C
Scope of effect exceeds scope of control

Modules Affected by Decision

Modula A is scope of control = A,B,C
Module C scope of control = C

Scope of effect should be limited to the module in which the decision is made and its immediate sub ordance

Figure7: Illustration of scope of effect versus scope of control

7.3 Object-oriented design.

Numerous SED methods based on objects have been proposed. The field has evolved from the early object-based design of the mid-1980s (noun = object; verb = method; adjective = attribute) through object-oriented design, where inheritance and polymorphism play key roles. *Object-oriented design* stems from the concept of *data abstraction*; *responsibility-driven design* has also been proposed as an alternative to object-oriented design [SWEBOK 2004; http://en .wikipedia.org/wiki/Object-oriented_programming].

(1) **Encapsulation** — *Encapsulation* refers to the creation of self-contained modules that bind processing functions to the data. These user-defined data types are called "classes," and one instance of a class is an "object." Encapsulation ensures good code modularity, which keeps routines separate and less prone to conflict.

(2) **Inheritance** — The purpose of *inheritance* is to pass "knowledge" down. Classes are created in hierarchies, and inheritance lets the structure and methods in one class pass down the hierarchy to another class. That means less programming is required when adding functions to complex systems. If a step is added at the bottom of a hierarchy, only the processing and data associated with that unique step must be added. Everything else above that step is inherited. The ability to reuse existing objects is considered a major advantage of object technology.

(3) **Polymorphism** — *Object-oriented programming* lets programmers create procedures for objects whose exact type is not known until run-time. For example, a screen cursor may change its shape from an arrow to a line depending on the program mode. The routine to move the cursor on-screen in response to mouse movement can be written for "cursor," and polymorphism lets that cursor mimic simulating system behavior.

(4) **Data abstraction** — *Abstraction* involves the facility to define objects that represent abstract "actors" that can perform work, report on and change their state, and "communicate" with other objects in the system. The term "encapsulation" refers to the hiding of state details, but extending the concept of *data type* from earlier programming languages to behavior associated most strongly with the data, and standardizing the way that different data types interact, is the beginning of abstraction.

When abstraction proceeds into the operations defined, enabling objects of different types to be substituted, it is called "polymorphism." When it proceeds in the opposite direction, inside the types or classes, structuring them to simplify a complex set of relationships, it is called "delegation" or "inheritance."

7.4 Responsibility-driven design.

Responsibility-driven design is a design technique in object-oriented programming. It was proposed by Rebecca Wirfs-Brock and Brian Wilkerson [1990], who defined it as follows:

(1) What actions make this approach responsible?

(2) What information does this object share?

Responsibility-driven design is in direct contrast to *data-driven design*, which promotes defining class behavior along the data that it holds. Data-driven design is not the same as data-driven programming, which is concerned with using data to determine control flow, not class [http://en.wikipedia.org/wiki.Resonsibility-driven_design].

7.5 Data-structure-centered design.

An example of *data-structure-centered design* is the *Jackson System Development (JSD)* model, which starts from the data structures rather than from the function. The software engineer first describes the input and output data structures and then develops the program's control structure based on these data structure diagrams. Various heuristics have been proposed to deal with special cases—for example, when there is a mismatch between the input and output structures.

The Jackson System Development (JSD) is a method of system development that covers the software life cycle either directly or by providing a framework into which more specialized techniques can fit. JSD can start in a project when there is only a general statement of requirements.

However, many projects that have used JSD actually started slightly later in the life cycle, doing the first steps largely from existing documents rather than directly with the users. The later steps of JSD produce the code of the final system. Jackson's first method, *Jackson Structured Programming (JSP)*, is used to produce the final code. The outputs of the earlier steps of JSD are a set of program design problems, the design of which is the subject matter of JSP.

Maintenance is also addressed by reworking appropriate earlier steps [http://en.wikipedia.org/wiki/jackson/development]. Three basic principles of operation of JSD are as follows:

(1) *Development* must start with describing and modelling the real world, rather than specifying or structuring the function performed by the system.

(2) A system made using the *JSD method* performs the simulation of the real world to the function or purpose of the system.

(3) An *adequate model* of a time-ordered world must itself be time ordered. The primary aim is to map progress in the real world onto progress in the system that models it.

7.6 Component-based design (CBD).

A *software component* is an independent unit having well-defined interfaces and dependencies that can be composed and deployed independently. Component-based design addresses issues related to developing, and integrating such components in order to improve reuse [SWEBOK 2004].

Note that *reuse* and *off-the-shelf software* need to meet the same security and reliability requirements as new software.

7.7 Other methods.

Other interesting but less mainstream approaches also exist; for example, formal and rigorous approaches: *SADT, RUP, state charts, VDM,* and *Z. Aspect-oriented design* is a method by which software is constructed using aspects to implement the crosscutting concerns and extensions that are identified during the software requirements process. *Service-oriented architecture* is a way to build distributed software using web services executed on distributed computers.

Software systems are often constructed by using services from different providers because standard protocols (such as *HTTP, HTTPS,* and *SOAP*) have been designed to support service communication and service information exchange.

(1) **VDM** — The *Vienna Development Method (VDM)* is one of the longest-established formal methods for the development of computer-based systems. Originating in work done at IBM's Vienna Laboratory in the 1970s, it has grown to include a group of techniques and tools based on a formal specification language—the VDM Specification Language (VDM-SL). It has an extended form, VDM++ that supports the modeling of object-oriented and concurrent systems.

Support for VDM includes commercial and academic tools for analyzing models, including support for testing and proving properties of models and generating program code from validated VDM models. There is a history of industrial usage of VDM and its tools, and a growing body of research in the formalism has led to notable contributions to the engineering of critical systems, compilers, concurrent systems, and logic for computer science [http://en.wikipedia.org/wiki/Vienna_Development_Method].

(2) **SADT** — The *Structured Analysis and Design Technique (SADT)* is a systems engineering and software engineering methodology for describing systems as a hierarchy of functions. SADT is a diagrammatic notation designed specifically to help people describe and understand systems. It offers building blocks to represent entities and activities, and a variety of arrows to connect boxes.

These boxes and arrows have associated informal semantics. SADT can be used as a functional tool for analyzing a given process using successive levels of details. The SADT method not only allows one to define user needs for information technology (IT) developments, which is often used in industrial information systems, but also to explain and present an activity's manufacturing processes and procedures.

The SADT supplies a specific functional view of any enterprise by describing the functions and their relationships in a company. These functions fulfill the objectives of a company, such as sales, order planning, product design, part manufacturing, and human resource management. The SADT can depict functional relationships here and can reflect data-control and control-flow relationships between different functions [http://en.wikipedia.org/wiki/Structured_Analysis_and_Design_Tech nique]. *See also Chapter 2, Paragraph 3.3.2.*

(3) **RUP** — The *Rational Unified Process (RUP)* is an iterative software development process framework created by the Rational Software Corporation, a division of IBM since 2003. RUP is not a single concrete prescriptive process, but rather an adaptable process framework, intended to be tailored by the development organizations and software project teams that will select the elements of the process that are appropriate for their needs. RUP is a specific implementation of the Unified Process [http://en.wikipedia.org/wiki/Rational_Unified_Process].

(4) **State charts** — A s*tate chart* is a type of diagram used in computer science and related fields to describe the behavior of systems. State charts require that the system described be composed of a finite number of states; sometimes this is indeed the case, while at other times this is a reasonable abstraction. Many forms of state charts exist (such as HTTP, HTTPS and SOAP), which differ slightly and have different semantics.

(5) **HTTP** — The *Hypertext Transfer Protocol (HTTP)* is an application protocol for distributed, collaborative, and hypermedia information systems. HTTP is the foundation of data communication for the World Wide Web. Hypertext is structured text that uses logical links (hyperlinks) between nodes containing text. HTTP is the protocol to exchange or transfer hypertext [http://www.w3.org/Protocols/].

(6) **HTTPs** — The *Hypertext Transfer Protocol Secure (HTTPS)* is a communications protocol for secure communication over a computer network, especially on the Internet. Technically, HTTPS is not a protocol in and of itself; rather, it is the result of simply layering the Hypertext Transfer Protocol (HTTP) on top of the Secure Sockets Layer/Transport Layer Security (SSL/TLS) protocol, thus adding the security capabilities of

SSL/TLS to standard HTTP communications. The main motivation for HTTPS is to prevent wiretapping and other attacks [http://en.wikipedia .org/wiki/HTTP].

(7) **SOAP** — Originally termed *Simple Object Access Protocol,* SOAP is a protocol specification for exchanging structured information in the implementation of web services in computer networks. It relies on the XML Information Set for its message format, and it usually relies on other application layer protocols, most notably Hypertext Transfer Protocol (HTTP) or Simple Mail Transfer Protocol (SMTP), for message negotiation and transmission [http://en.wikipedia.org/wiki/SOAP].

8. Design Tools

Software design tools can be used to support the creation of SED artifacts during the software development process [SWEBOK 2014]. Software analysis and design includes all activities, which help to implement the transformation of requirements specifications. Requirements specifications specify all functional and non-functional expectations from the software. These requirements specifications are created as human readable and understandable documents, independent of computers.

Software analysis and design is the intermediate stage, which helps human-readable requirements to be transformed into actual code. Some examples are [http://www.tutorialspoint.com/software_engineering/software_analysis_desig n_tools.htm]:

(1) **Data dictionary** — A *data dictionary* is the centralized collection of information about data. It stores meaning and origin of data, its relationships with other data, data format for usage, etc. A data dictionary has rigorous definitions of all names in order to facilitate user and software designers.

(2) **Data flow diagram** — A *data flow diagram* is a graphical representation of flow of data in an information system. It is capable of depicting incoming data flow, outgoing data flow, and stored data.

(3) **Decision tables** — A *decision table represents* conditions and the respective actions to be taken to address them, in a structured tabular format.

(4) **Entity-relationship model** — An *entity-relationship model* is a type of database model based on the notion of real world entities and relationships among them.

(5) **HIPO diagram** — A *HIPO (hierarchical input process output) diagram* is a combination of two organized methods to analyze the system and pro-

vide the means of documentation. The HIPO model was developed by IBM in 1970.

(6) **Pseudo-code** — *Pseudo code* is written more closely to the selected programming language. It may be considered as an augmented programming language, full of comments and descriptions.

(7) **Structure charts** — A *structure chart* is a chart derived from a data flow diagram. A structure chart represents the hierarchical structure of modules. A specific task is performed at each layer.

(8) **Structured English** — *Structured English* uses plain English words in a structured programming paradigm. It is not the ultimate code but a description detailing is necessary to code and how to code it.

Appendix A
Architectural Styles

An *architectural style* is "a set of constraints on an architecture that defines a set or family of architectures that satisfies them" [Bass, Clements and Kazman 2003]. An architectural style can thus be seen as a meta-model, which can provide software's high-level organization (its macro-architecture) [SWEBOK 2004]. The grouping of the various architectural styles varies with the numerous sources of information about architectural styles—usually a software textbook. Two popular *architectural style* books are Garland & Shaw [1994] and Sommerville [2011].

(1) **General structures.** *See layers, pipes, filters, and blackboard architectures.*

   a. **Layered architecture.** A *layered architecture* organizes the system into layers with related functionality associated with each layer. A layer provides services to the layer above it, so the lowest-level layers represent core services that are likely to be used throughout the system. An example of a layered model is a system for sharing copyright documents held in different libraries.

   Sommerville [2011] points out that this layered approach supports the incremental development of systems. As a layer is developed, some of the services provided by that layer may be made available to users.

   The architecture is also changeable and portable. So long as its interface is unchanged, another equivalent layer can replace a layer. Furthermore, when layer interfaces change or new facilities are added to a layer, only the adjacent layer is affected.

   As layered systems localize machine dependencies in inner layers, it is easier to provide multiplatform implementations of an application system. Only the inner, machine-dependent layers need be reimplemented to take account of the facilities of a different operating system or database.

There are several *advantages* of a layered system. It allows replacement of entire layers so long as the interface is maintained. Redundant facilities (e.g., authentication) can be provided in each layer to increase the dependability of the system.

In practice, the *disadvantage* of providing a clean separation between layers is that often it is difficult and at a high-level layer that may have to interact directly with lower-level layers rather than through the layer immediately below it. Performance can be a problem because of the multiple levels of interpretation of a service request as it is processed at each layer [Sommerville 2011].

b. ***Pipe-and-filters (stovepipe) architecture.*** In *pipe-and-filter* style, each component has a set of inputs and a set of outputs. A component reads streams of data on its inputs and produces streams of data on its outputs. This is usually accomplished by applying a logical transformation to the input streams and computing incrementally, so that output begins before input can be completed.

Hence, components are termed "filters." The connectors of this style serve as conduits for the stream, transmitting outputs of one filter to inputs of another. Hence, the connectors are termed "pipes" [Garlan & Shaw 1994].

The *advantages* of pipe-and-filter architecture are that this approach is easy to understand and it supports transformation reuse. Workflow style matches the structure of many business processes. Evolution by adding transformations is straightforward and can be implemented as either a sequential or a concurrent system. A *disadvantage* is that each transformation must parse its input and un-parse its output to the agreed form. This increases system overhead and may mean that it is impossible to reuse functional transformations that use incompatible data structures [Sommerville 2011].

c. ***Blackboard architecture.*** A *blackboard* is a place where information is posted. It only knows about itself, not about its observers. Each observer watches the blackboard via polling. When the blackboard is updated, each observer decides for itself if it needs to respond to the update. Observers do not know about each other; they only know about the blackboard.

The blackboard's *strengths* are that observer applications can be added or deleted at any time without affecting the application. In contrast, its *weakness* can be considerable overhead for the observer's applications to poll the blackboard [Garlan & Shaw 1994].

(2) **Distributed systems.** *See client-server (two-tier), and three-tier system, and broker architectures.*

    a. **Client-server architecture.** A *client-server system* is also called a "two-tier system" or a "layered system." A layered system is organized hierarchically, each layer providing service to the layer above it and serving as a client to the layer below. In some layered systems, inner layers are hidden from all except the adjacent layer, except for certain functions carefully selected for export (i.e., when system components implement a *virtual machine* at some layer in the hierarchy). The connectors are defined by the protocols that determine how the layers will interact.

    There are two client-server types [Garlan & Shaw 1994]:

    o **Fat client** — A "fat client" has knowledge of the information and can perform information manipulation. The level of manipulation depends upon how "smart/fat" the client is.

    o **Thin client** — A "thin client" has no "brains." The client accepts information from the server and displays the information without any manipulation. A Web is an example of a thin client.

    b. **Three-tier architecture.** *Three-tier architecture* is also an example of a layered system—in this case, three layers. The first layer houses the HCI. The second/middle layer houses the knowledge/rules of the system, and the third layer houses any specialty systems like a database.

    Most systems today are at least a three-tier system. The top layers isolate (screen) the "user" from the "real" data.

    The middle layers hold all the knowledge/rules of the system (how to manipulate the data from the storage/database into user information for the screens). The bottom layers hold the actual data or specialty systems.

    The database is a good example of a specialty system. It should not be necessary for the middle layer to know the format of the information it needs. The middle layer requests data and the database knows how to both retrieve the data and return it.

The *strengths* of the three-tier architecture system involve the *presentation layer*, which is separated from the database by the application tier. It is easy to change the look and feel of the database with relatively minor effects on the rest of the application. This leads to a consistent user interface across the whole application.

The *weaknesses* of the three-tier pattern are that usually some kind of transaction management service is required to track transactions from presentation tier to database, and changes in functionality typically require changes to all three tiers of the architecture [Garlan & Shaw 1994].

c. **Broker architecture.** The *Common Object Request Broker Architecture (CORBA)* is a *specification* of a standard architecture for object request brokers (ORBs). A standard architecture allows vendors to develop ORB products that support application portability and interoperability across different programming languages, hardware platforms, operating systems, and ORB implementations.

In using a CORBA-compliant ORB, a client can transparently invoke a method on a server object, which can be on the same machine or across a network. The ORB intercepts the call and is responsible for finding an object that can implement the request, passing it the parameters, invoking its method, and returning the results of the invocation. The client does not have to be aware of the object's location, its programming language, its operating system, or any other aspects that are not part of the interface [http://www.omg.org (1996)].

(3) **Interactive systems.** An *interactive system* is designed with model-view–controller or presentation-abstract-controller architecture. (*See model-view-controller and P\presentation-abstract-controller architectures.*)

a. **Model-view-controller (MVC) architecture.** The *MVC architecture* divides a given software application into three interconnected parts in order to separate internal representations from the ways that information is presented to or accepted from the user. The notions of separation and independence are fundamental to architectural design because they allow changes to be localized.

The *MVC pattern* separates elements of a system, allowing them to change independently. For example, adding a new view or changing an existing view can be done without any changes to the underlying data in the model. A layered architecture is another way of achieving separation and independence.

A *view* can be any output representation of information, such as a chart or a diagram. Multiple views of the same information are possible—a bar chart for management and a tabular view for accountants. The third part, the controller, accepts input and converts it to commands for the model [http://en.wikipedia.org/wiki/Model–view–controller].

The *advantages* of MVC architecture are that it allows the data to change independently of its representation and vice-versa. This architecture supports presentation of the same data in different ways. The *disadvantages* are that this architecture can involve additional code and code complexity even when the data model and interactions are simple [Sommerville 2011].

b. **Presentation-abstract-controller (PAC) architecture.** The *PAC architecture* is another architectural pattern. The PAC is somewhat similar to MVC in that it separates an interactive system into three types of components responsible for specific aspects of the application's functionality.

The abstraction component retrieves and processes the data, the presentation component formats the visual and audio presentation of data, and the control component handles things such as the flow of control and communication between the other two components.

In contrast to MVC, PAC uses a hierarchical structure of agents, each consisting of a triad of presentation, abstraction, and control parts. The agents (or "triads") communicate with each other only through the control part of each triad. PAC also differs from MVC in that within each triad, it completely insulates the presentation ("view" in MVC) and the abstraction ("model" in MVC). This provides the option to separately multithread the model and view, which can allow the user to experience very short program start times, as the user interface (presentation) can be shown before the abstraction has fully initialized [http://en.wikipedia.org/wiki/Presentation%E2%80%93abstraction%E2%80%93control].

(4) **Adaptable (interactive) systems.** Two examples of *adaptive systems* are *microkernel architecture* and *reflection architecture*.

a. **Microkernel architecture.** A *microkernel* (also known as a μ-kernel or Samuel kernel) is the near-minimum amount of software that can provide the mechanisms needed to implement an operating system. These mechanisms include low-level address space management, thread management, and interprocess communication (IPC). If the hardware provides multiple rings or CPU modes, the microkernel is the only software executing at the most privileged level (generally referred to as supervisor or kernel mode).

Traditional operating systems, such as device drivers, protocol stacks, and file systems, are removed from the microkernel to run in user space. In source code size, microkernels generally tend to be under 10,000 lines of code. A MINIX's kernel, for example, has fewer than 6,000 lines of code [http://en.wikipedia.org/wiki/Microker nel].

   b. ***Reflection architecture.*** The *reflection architectural* pattern provides a mechanism for changing the structure and behavior of software systems dynamically. It supports the modification of fundamental aspects, such as type structures and function call mechanisms. In this pattern, an application is split into two parts. A *meta level* provides information about selected system properties and makes the software self-aware. A *base level* includes the application logic. Its implementation builds on the meta level. Changes to information kept in the meta level affect subsequent base-level behavior [http://wiki.ifs.hsr.ch/APF/files/Reflection.pdf.].

(5) ***Other systems.*** *See batch, interpreters, process control, and rule-based architectures.*

   a. ***Batch architecture*** —The 1950s were marked by the development of rudimentary operating systems designed to smooth the transitions between jobs (a job is any program or part of a program that is to be processed as a unit by a computer). This was the start of *batch processing*— programs to be executed were grouped into batches.

   While a particular program was running, it had total control of the computer. When it finished, however, control was returned to the operating system, which handled any necessary finalizations, read, and started up the next job. Since computers could easily handle the transition between two jobs instead of having it done manually, less time was taken, and the CPU was more efficiently utilized [Slotnick et al. 1986].

   b. ***Interpreter architecture*** — *Interpreters* are commonly used to build virtual machines that close the gap between the computing engine expected by the semantics of the program and the computing engine available in hardware. For example, compilers could use an interpreter to generate code [https://en.wikipedia.org/wiki/Inter preter_(computing)].

c. ***Process control architecture*** — This system organization is not widely recognized; nevertheless, it seems to appear within other designs. *Control-loop design* is characterized by both the kinds of components involved and the special relations that must hold among them. Continuous processes of many kinds convert input materials to products with specific properties by performing operations on the inputs and on intermediate products. The purpose of a control system is to maintain specific properties of the outputs of the process at given reference values (set points). Examples include:

  o ***Open-loop system*** — If the input materials are pure, if the process is fully defined, and if the operations are completely repeatable, the process can run without surveillance. Example: A hot-air furnace that uses a constant burner setting to raise the temperature of the air that passes through it.

  o ***Close-loop system*** — Most often, properties such as temperature, pressure, and flow rates are monitored, and their values are used to control the process by changing the settings of apparatus. Example: heaters, values, and chillers with a thermostat.

d. ***Rule-based architecture*** — *Rule-based systems* automate problem-solving knowledge, provide a means for capturing and refining human expertise, and prove to be commercially viable. Example: systems used in artificial intelligence [https://en.wikipedia.org/wiki/Rule-based_system].

REFERENCES

- **[ACM SIGCHI 1996]** Thomas T. Hewett, Ronald Baecker, Tom Carey, Jean Gasen, Marilyn Mantei, Gary Perlman, Gary Strong and William Verplank, *Curricula for Human-Computer Interaction*, ACM, New York, 1996.

- **[Alexander et al. 1977]** C. Alexander, S. Ishikawa, M. Silverstein, M. Jacobson, I. Fiksdahl-King and S. Angel, *A Pattern Language*, Oxford University Press, Oxford, England, 1977.

- **[Bass, Clements, & Kazman 2003]** L. Bass, P. Clements, and R. Kazman, *Software Architecture in Practice,* 2nd ed., Addison-Wesley, Reading, MA, 2003.

- **[Bernard, Liao, & Mills 2001]** M. Bernard, C.H. Liao, and M. Mills. "The Effects of Font Type and Size on the Legibility and Reading Time of Online Text by Older Adults," *Department of Psychology*, Wichita State University, Wichita, KS, 2001.

- **[Booch, Rumbaugh, & Jacobson 1999]** G. Booch, J. Rumbaugh, and I. Jacobson, *The Unified Modeling Language User Guide,* Addison-Wesley, Reading, MA, 1999.

- **[Bosch 2000]** J. Bosch, *Design & Use of Software Architectures: Adopting and Evolving a Product-Line Approach,* 1st ed., ACM Press, New York, 2000.

- **[Budgen 2003]** David Budgen, *Software Design*, 2nd ed., Pearson, Addison-Wesley, Reading, MA, 2003. (Recommended as an IEEE PSEM Certification exam reference by the IEEE Computer Society.)

- **[Buschmann 1996]** F. Buschmann, *Pattern-Oriented Software Architecture: A System of Patterns,* John Wiley & Sons, New York, 1996.

- **[Christensen 2005]** Mark Christensen, "Implementing and Testing the Design," in *Software Engineering: The Development Process*, IEEE Computer Society Press, Los Alamitos, CA, 2005.

- **[Clements et al. 2002]** Paul Clements, Felix Bachmann, Len Bass and David Garlan, *Documenting Software Architectures: Views and Beyond,* Addison-Wesley, Reading, MA, 2002. (Recommended as an IEEE PSEM Certification exam reference by the IEEE Computer Society.)

- **[DeMarco 1999]** Tom DeMarco, "The Paradox of Software Architecture and Design," *Stevens Prize Lecture,* August 1999.

- **[Draper 1998]** Steve W. Draper, "Computer Supported Cooperative Lecture Notes,"University of Glasgow, Glasgow, Scotland, 1998.

- **[Dumais & Czerwinski 2001]** Susan Dumais and Mary Czerwinski, "Building Bridges from Theory to Practice." in *Microsoft Research,* One Microsoft Way, Redmond, WA, 2001.

- **[Shaw and Garlan 1996]** M. Shaw and D. Garlan, *Software Architecture: Perspectives on an Emerging Discipline*, Prentice-Hall, Upper Saddle River, NJ, 1996.

- **[IEEE 1016-1998]** *IEEE Standard 1016-1998*, "IEEE Recommended Practice for Software Design Descriptions, IEEE Computer Society, NY, 1998.

- **[Kiczales et al. 1997]** G Kiczales Gregor Kiczales, John Lamping, Anurag Mendhekar, Chris Maeda, and Cristina Videira, *Aspect Oriented Programming—UBC Computing Science,* Springer-Verlag, Berlin, 1997.

- **[Liskov & Guttag 2001]** B. Liskov and J. Guttag, *Program Development in Java: Abstraction, Specification, and Object-Oriented Design*, Addison-Wesley, Reading, MA, 2001.

- **[Riel and Webber 1984]** H.J. Riel and M.M. Webber, "Planning Problems are Wicked Problems," N. Cross (eds.), *Developments in Design Methodology*, Wiley, Hoboken, NJ, 1984, pp. 135-144.

- **[Sommerville 2007]** Ian Sommerville, *Software Engineering*, 8th ed., Addison-Wesley, Harlow, England, 2007. (Recommended as an IEEE PSEM Certification exam reference by the IEEE Computer Society.)

- **[Sears & Jacko 2007]** Andrew Sears and Julie A. Jacko, *The Human-Computer Interaction Handbook: Fundamentals, Evolving Technologies and Emerging Applications*, 2nd ed., Taylor & Francis, Boca Raton, FL, 2007.

- **[Slotnick et al. 1986]** D.L. Slotnick, E.M. Butterfield, E.S. Colantonio, D.J. Kopetzky and J.K. Slotnick, *Computers & Applications with BASIC*, D.C. Heath and Company, Lexington, MA, 1986.

- **[Sommerville 2007]** Ian Sommerville, *Software Engineering*, 8th ed., Addison-Wesley, Harlow, England, 2007. (Recommended as an IEEE PSEM Certification exam reference by the IEEE Computer Society.)

- **[SWEBOK 2004]** *Guide to the SWE Body of Knowledge*, IEEE, New York, 2004.

- **[Wikipedia]** Wikipedia is a free web-based encyclopedia enabling multiple users to freely add and edit online content. Definitions cited in Wikipedia and their related sources have been verified by the authors and other peer reviewers. Readers who would like to verify a source or a reference should search the subject on Google and read the technical report found under Wikipedia or another web site.

- **[Wirfs-Brock & Wilkerson 1990]** R. Wirfs-Brock and B. Wilkerson, *Designing Object-Oriented Software*, Prentice Hall, Upper Saddle River, NJ, 1990.

- **[Yourdon 1989]** Edward Yourdon, *Modern Structured Analysis*, Prentice Hall, Upper Saddle River, NJ, 1989.

Chapter 2

The Nature of Software Design[2]

David Budgen
University of Durham, U.K.

While software is almost all pervasive in the modern world, the act of designing software poses some very significant challenges. The aim of this overview paper is, therefore, to describe the key properties of software; to explain how these influence the design process; and to review some major examples of the strategies and forms that have evolved to address properties, design, strategies, and forms.

1. The Nature of Software Design

First, we should ask the question:

What exactly is the purpose of design?

The answer to this question essentially defines the scope of this paper by identifying the issues that any design solution should address. There are many possible answers, which will reflect the context when a particular design task is undertaken. However, a reasonably generic answer is:

> *To produce a plan (or model) that represents a workable*
> *(implementable) solution to a given need.*

In the context of SED, we are seeking to create a "solution" that involves creating some form of artifact from appropriate software forms. In the remainder of the first section, we therefore examine how SED is influenced by the nature of design activities, by the particular characteristics of software, by the context within which SED is performed, and by our ideas about what might constitute a "good" design.

The following sections then examine some of the conceptual tools that we employ to assist with SED. Section 2 discusses some of the different ways that software can be organized and broadly classified as *architecture*. Section 3 examines how we can *visualize* ideas about a design with different notations. Together, these then underpin Section 4, where we review some well-established ways of organizing the design *process*.

---

2. Based on "Software Design: An Introduction," by David Budgen, which appeared in R.H. Thayer and M. Dorfman (editors), *Software Engineering*, Volume I, 3rd edition, IEEE Computer Society Press, Los Alamitos, CA, ©2007, IEEE. Used with permission of the author.

1.1 The nature of designing.

Software design activities need to conform to the constraints imposed by the nature of designing in general. Design problems are widely recognized to be "wicked" problems [Rittel and Webber 1984].

A wicked problem (a.k.a. "ill-structured problems") is characterized by having such properties as:

- No *true/false* solutions, with many possible solutions, that can only be ranked as better or worse from a particular perspective.

- No definitive formulation so that the specification and understanding of a problem are bound with our ideas about "solving" it.

- No stopping rule that can be used to determine when an optimum design has been achieved.

- Immediate and ultimate tests that can be used to determine that a design solution fits the needs of the problem (i.e., the requirements).

A key consequence is that the activity of designing cannot ever be a "procedural" or "defined" process. Indeed, the design process is essentially *empirical* or *opportunistic* in nature [Hayes-Roth and Hayes-Roth 1979, Williams and Cockburn 2003] and involves exploration of a very large "solution space."

These characteristics can be illustrated by a very simple example of a design task that will be familiar to many people, which is that of moving to a new home. When we move to a new house or apartment, we are faced with a classical design problem: namely that of determining where our furniture is to be placed. Indeed, we may also be expected to provide the moving company with a design "plan" to indicate our intentions.

There are of course many ways in which furniture can be arranged within a house or apartment. For each item, we need to decide in which room we want it to be placed, perhaps determined mainly by functionality, and then where it might go within that room. We might choose to concentrate on getting a good balance of style and arrangement in more public rooms at the expense of others. The form of the building will also provide constraints, in that furniture should not block doors or windows and should leave power outlets and other connections accessible.

The process of designing software is not so very different from this, and exhibits the same forms and issues, further complicated by some properties of software itself that we now need to consider.

1.2 Characteristics of software.

Given that an important task for the software designer is to formulate some form of abstract design model that represents his or her ideas about a design solution, we might ask what causes this to be so great a challenge? Fred Brooks [1987] suggested software characteristics that contribute to this, including:

- *The complexity of software*, with no two parts ever being quite alike, needs a process having many possible states during execution.

- *The problem of conformity* arises from the very pliable nature of software; therefore, the designer is expected to tailor software to meet the needs of hardware, of existing systems or to meet other "standards."

- The (apparent*) ease of changeability of software* means that users are apt to expect changes to be made without an appreciation of the true costs (in terms of money, resources, and structure) that these imply.

- The *invisibility of software* means that our descriptions of our design ideas lack any clear visual link to the form of the product, and so are unable to help with comprehension of ideas in the same way that occurs with more "physically" connected forms of description.

Together these explain why it is so difficult to clearly and unambiguously "capture" any ideas that we might have about the design for a software system. Designing is always a challenging activity, and for software it is rendered even harder by these characteristics.

1.3 What constitutes design knowledge?

The process of learning about design may well involve both a period of academic study and a spell of "apprenticeship," which involves learning directly from a more experienced practitioner by working with them in some way. Regardless of how this may be organized, the aim is to provide a fledgling designer with both experience and an understanding of how to deploy the design elements available in the particular medium effectively.

Experience plays an important part in designing software that has been studied and is consistent with this view of design knowledge. Adelson and Soloway [1985] noted a range of techniques being used, with the choice being dependent both upon the expertise of the designer and familiarity with the given problem domain. They particularly noted the use of "labels for plans" by experts, whereby a designer identified a part of the task for which they could reuse prior design experience, "labelling" this intention at an abstract level.

A later study of expert designers by Guindon [1990] observed that a variety of *knowledge schemas* was employed, from simple rules to part solutions. For object-oriented development, Détienne [2002] noted the use of three forms of knowledge schema—*application domain schema, function schema,* and *procedure schema.* From a cognitive perspective, a *schema* can be considered as a form of internal "knowledge structure" that an expert employs to represent "generic concepts stored in memory," and it is their possession of a richer set of internal schema that largely distinguishes experts from those less experienced.

1.4 The software development life cycle.

Designing software is not an isolated and independent activity. The eventual system as implemented will be expected to meet a whole set of user needs, which reminded us of the criterion of "fitness for purpose." In a classical software life cycle such as the well-known "waterfall" model, it is expected that these needs will be determined in advance through some form of *requirements elicitation* process, possibly aided by an *analysis* of what the system is to do. However, in reality, these tasks are likely to interact, both with each other and with the activities of design, since each step can both constrain later steps and reveal inconsistencies in the earlier ones.

In addition, it is expected that the designer will provide a set of specifications for those whose job it is to construct the system. These will need to be clear, complete and unambiguous, yet despite this (if such an ideal can be achieved), it is likely that further needs for change will emerge during implementation. Furthermore, over and above such immediate issues, the designer also needs to think about the long-term evolution of a system and seek to devise a structure that can accommodate any likely changes.

The sheer difficulty of balancing this medley of conflicting goals has led to the emergence of a quite different way of thinking about the software development context that we describe as *Agile methods*. These seek to recognize the uncertainties in the overall development process, and assume that the form of the system will evolve as understanding of its role emerges. For such forms, the role of design becomes much more closely entwined with those of requirements elicitation and implementation. As far as design is concerned, an important aspect of such methods is to ensure that constant change and evolution do not undermine design qualities and structures.

1.5 Quality factors for design.

Quality can be an elusive concept at best, and given the properties of software discussed above, it is not surprising that this is particularly so for SED. Indeed, our quality concerns are usually involved with our particular relationship with the system itself.

Having suggested that the concept of "fitness for purpose" was a paramount goal for any system, we do not need to recognize that this cannot be directly measured, nor is it absolute. Simply performing the specified task(s) correctly and within the given resource constraints may not be enough to achieve fitness for purpose. An example of this would be a system that is expected to be in service for at least ten years with modification at frequent intervals.

For this case, our notions of fitness for purpose are very likely to incorporate ideas about how well the overall structure can be adapted to accommodate the likely changes without compromising the other qualities. The converse is equally true. Where a system is urgently required to meet a short-term need, getting a system that works (correctly) will be more important than ensuring that it can be modified and extended.

Space limits what we can say here about quality factors, but a useful group to note is those that are generally referred to as the "*ilities*." The exact composition of this group may be dependent upon context, but the key ones are generally those of *efficiency*, *reliability*, *maintainability*, and *usability*. They describe rather abstract "top-level" properties of the eventual system that cannot be easily accessed from design information alone.

Indeed, devising suitable ways to measure design information in a reliable and systematic manner is something of a challenge. While at the level of implementation we can employ basic code measurement (metrics) by such means as counting lexical tokens [Fenton and Pfleeger 1997], the variability and the weak syntax and semantics associated with design notations make such an approach less appropriate for design. Hence, more qualitative forms such as design walkthroughs and design reviews may well be more suitable [Parnas and Weiss 1987].

2. Software Architecture

The idea of *architecture* in connection with software began to emerge in the early 1990s. To some extent, this probably reflected a growth in the different ways of organizing software systems. Where once almost all software was organized based on a main program unit invoking a set of subprograms (what we now usually term "call-and-return"), later systems began to be organized around other forms such as objects, processes that were spread across a range of different computers and large databases.

Various terms were used to capture these ideas, and as usual when something new emerges, these were not always used consistently, thus reducing their value as a "vocabulary." In the next subsection, we will briefly look at some examples of what we now term *software architecture*.

2.1 Basic concepts.

An early and very clear discussion of this appeared in a paper by David Garlan and Dewayne Perry [1995], written as an introduction to a collection of papers on this topic. In this work, they examined some of the roles that the concept could perform, including: helping with understanding of a high-level design through the provision of an abstract vocabulary; helping to identify where elements could be reused; and providing an understanding of "a system is expected to evolve." Indeed, in many ways, the architecture of a system is simply the abstract form of its top-level design.

The book by Mary Shaw and David Garlan [1996] provided a valuable baseline for the emerging ideas. In this, they employed a basic framework of describing an architecture in terms of the kinds of *components* and *connectors* employed in a given system architecture. Their book examined and classified a number of *architectural styles* based upon these ideas, and hence, had the benefit of clarifying a vocabulary that was increasingly being used, but not always consistently.

Ideas about architecture and about its influence upon such developments as the concept of "software product lines" involving the reuse of architectural ideas for multiple systems have continued to evolve. For a fuller understanding of this area, a book such as Taylor, Medvidovic, and Dashofy [2010] provides more detail as well as examples. For our purposes though, the basic concepts should be a sufficient introduction.

2.2 Some architectural styles.

The concept of an *architectural style* has proved to be a useful one in a number of other domains. (*See also Paragraph 2.3*) When speaking of buildings, referring to a house as being in a "black and white" style tells us about its likely characteristics—with external wooden beams and small windows. The same concept applies to ships where the term "naval architecture" has long been employed and where it is recognized that the overall characteristics of a ship will reflect its purpose. Aircraft carriers have large, flat decks and a small superstructure to one side; oil tankers have large tanks in the main hull and small superstructures at the stern, and so forth. In the case of buildings, style may be dictated partly by the materials available during the period when it was constructed; for ships, it is driven more by function.

Software architecture is probably driven by all of these same approaches. The main influences being the type of elements used the way that they interact, and the purpose of the system. Table 1 below summarizes some common examples of software architectural styles, drawing upon the categorizations proposed in Shaw and Garlan [1996].

2.3 Architectural patterns.

While the notion of architectural style tells us something about the type of elements within a system and how they interact, an *architectural pattern* focuses more upon the overall organization of the elements. A useful introduction with illustrations is provided in Buschmann, et al. [1996]. Here, we illustrate the concept by discussing two particularly familiar forms.

2.3.1 Model-View-Controller (MVC). This is a widely used form in which the overall design of an interactive application is organized as three elements, each with a clearly defined roles and functionality:

- The model contains the core functionality and any relevant data.

- Views provide information to the user.

- Controllers handle user input. Each view has an associated controller that also handles related forms of input.

Table 1: Some examples of architectural style

Category	Characteristic	Example of Style
• Data-flow	Movement of data with recipients having no control of contents.	Data sequential Pipe-and filter
• Call-and-return	Single thread of control determining order of completion.	Main program or subprograms Classical objects
• Data-centered repository	Focus upon a complex control data-store.	Transactional data-bases Client-server Blackboard

The *user interface* then consists of views and controllers together and is independent of the model itself. However, the model will need to propagate information about changes to the controllers. Such an approach makes it easy to change the interface for a new platform or to employ new forms for presenting the information. Figure 1 illustrates the MVC structure.

2.3.2 Layers. The *layers* pattern is another form that is employed in many roles. It is commonly (but not exclusively) used for organizing a hierarchy of protocols (such as that of the OSI seven-layer model used in computer networking). The idea is that each layer will deal with a specific aspect of communication and will employ the services of the layer below (and no other layers) while providing service to the layer above.

The value of this pattern is that specific functionality related to a particular layer of abstraction is encapsulated in a layer and may easily be redeployed into a new context simply by substituting new layers below it. This form is illustrated in Figure 2, where we use part of the network OSI model to illustrate the characteristics.

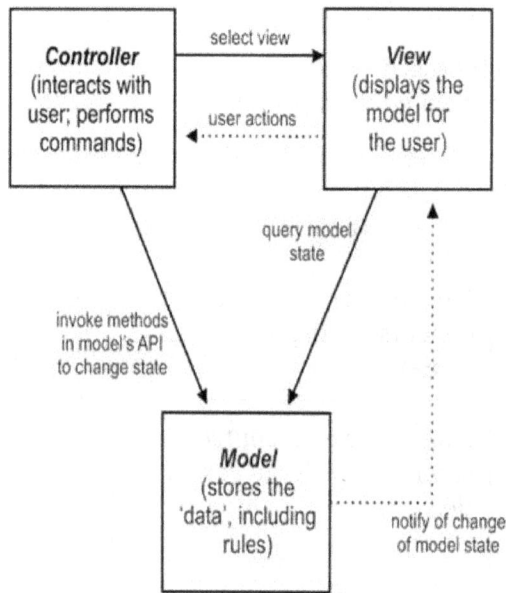

Figure 1: The Model-View-Controller pattern

3. Describing Designs

When discussing the nature of software, it was observed that its characteristics, and particularly that of *invisibility*, provide a challenge to any attempts to visualize our ideas about a design. However, regardless of the different roles for design notations explored in this section, we need them simply to avoid *cognitive overload* in developing design ideas. There is a limit to how much we can reliably "store" in our own working memory, and therefore, storage is limited to anything but the smallest problems. We simply need to find ways of visualizing design ideas even if they are not embedded in any form of "physical reality."

Over the years, software engineers have therefore developed a range of "box and line" notations intended to help with this task. Whether this has really had the attention (as a design task in itself) that it should perhaps have had is a moot point. Analyzed against a cognitive framework, most software engineering notations do seem to produce poor results [Moody 2009]. Anecdotally, experienced designers seem to produce informal diagrams to help develop their ideas, only reverting to notations that are more formal when these ideas need to be recorded and shared with others.

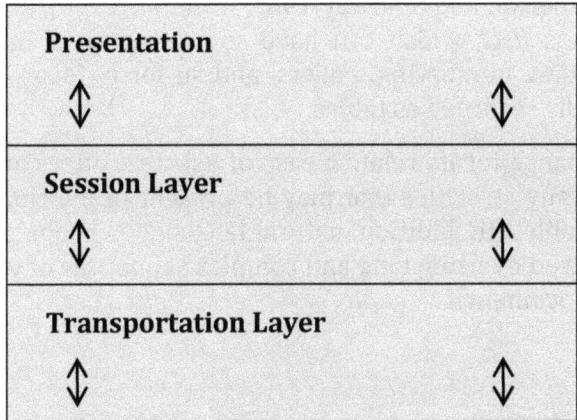

Presentation	
Session Layer	
Transportation Layer	

Figure 2: Example of a layer pattern

However, regardless of these issues, we still have a need for well-defined notations for such purposes as:

- Documenting and exploring our ideas about a design solution.

- Explaining our ideas to others (the customer, implementers, and other members of a design team).

- Checking for consistency and completeness of a design model.

In the next section, we examine a general categorization of design notations with some examples.

3.1 Design viewpoints.

The wide range of notational forms that we use can be categorized in a number of different ways. Some, such as the "4+1" model [Kruchten 1994] are closely linked to a particular architectural style (in this case, object-orientation). For this section though, we will employ a more generic categorization into four different groupings, described more fully in Budgen [2003].

This grouping is based upon the idea of a *design viewpoint*, where a viewpoint is considered as being a "projection" from the "internal" design model that displays certain characteristics with an appropriate level of abstraction.

Figure 3 provides a schematic illustration of this idea along with some examples of forms used to record these viewpoints.

The term "system" in the above definitions is used loosely, since at different times we might want to describe a complete design solution, or parts of it.

3.1.1 Text. *Text* is widely used, largely in conjunction with the other forms, but also on its own. The practice of *note-making* has been widely observed in studies of SED [Adelson and Soloway 1985, Guindon 1990], and such notes are often organized as *lists*, which can have some degree of informal structure through indentation, numbering, bullets, and so forth. Ideas and descriptions can also be usefully recorded as tables.

However, in exchange for its relative ease of use, text offers only limited scope for representing any structure that may be present in information beyond the use of lists and tables. In addition, natural language is prone to ambiguity that can only be resolved by using long and complex sequences of words (a good example is a legal document).

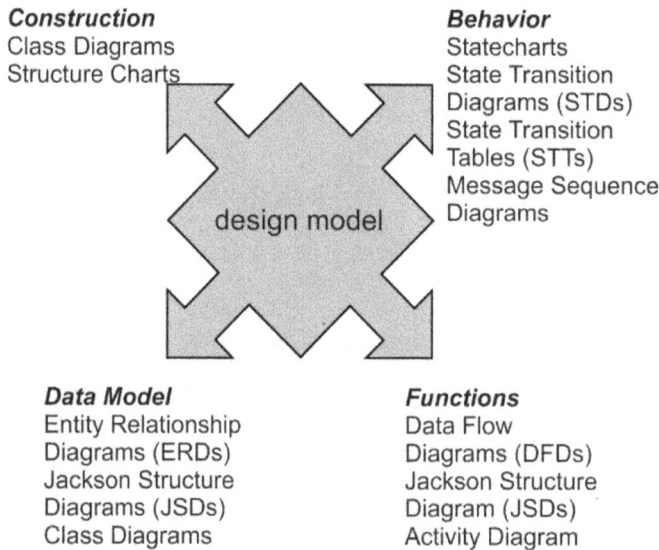

Construction
Class Diagrams
Structure Charts

Behavior
Statecharts
State Transition
Diagrams (STDs)
State Transition
Tables (STTs)
Message Sequence
Diagrams

design model

Data Model
Entity Relationship
Diagrams (ERDs)
Jackson Structure
Diagrams (JSDs)
Class Diagrams

Functions
Data Flow
Diagrams (DFDs)
Jackson Structure
Diagram (JSDs)
Activity Diagram

Figure 3: Examples of the design viewpoints

3.1.2 Diagrams. Since our examples will largely focus upon diagrams, we briefly describe two characteristics that appear to be significant for the successful use of diagrams.

The first relates to the number of symbols in use to describe the concepts (the "elements") of a diagram. As a loose rule of thumb, the more abstract the diagram the fewer the symbols. Diagrams with a large number of symbols tend to be more complex to use. (In this context, symbols can be many shapes or characters; they might be arrowheads, solid or dashed lines, and so forth.). A supplementary aspect of this is that we should also be able to draw the symbols easily—many designs are worked out and explored using whiteboards or pencil and paper and the designer wants to be able to concentrate on exploring an idea without needing to spend time drawing complex-shapes. Therefore, these symbols should be simple in form and ways by which they can be easily distinguished from each other.

The second is concerned with having a *hierarchy* within a notation. Large diagrams may be very difficult to understand, and many forms therefore allow the use of a hierarchy, whereby symbols at a higher level of abstraction are expanded into a "tree" of diagrams, with each level providing detail. Figure 4 demonstrates this idea in a schematic manner.

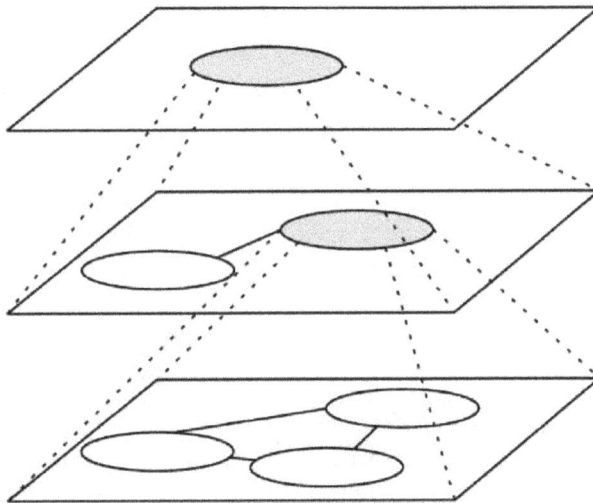

Figure 4: Hierarchy in representation

3.1.3 Mathematics. *Mathematical notations* are of course ideally suited to providing concise abstractions, so it is hardly surprising that they have been employed for this purpose using *Formal Description Techniques*, or DFDs for short. Traditionally, they are usually employed for the purpose of specification, whether this is due to system properties (for analysis and requirements specifications), or to the behavior and functionality of individual design elements. As with the case of text, it can be argued that these notations are most valuable when being used to support other descriptive forms, rather than being used solely on their own. On the downside, their use requires learning a set of (usually non-intuitive) symbols, and they are less well suited to describing larger-scale systems.

3.2 More examples.

Before discussing some simple examples of the concepts outlined above, we should note that the old saying about fire, that it "makes a good servant but a bad master" applies equally to the use of diagrams when designing software.

Any given form of diagram will have an established syntax (how we draw it) and semantics (what is meant by the symbols, their positioning and so forth.). However, the aim of a diagram is to *assist* with the process of design—too slavish a regard for syntax in particular during the evolution of design ideas can be a handicap. Indeed, observation suggests that experienced designers often produce informal diagrams while developing their ideas and then formalize these later. This is a point that we will return to when considering the use of tools to support the design process.

Below, we briefly describe how the different viewpoints are used within different system forms. We will also provide some simple examples of how these viewpoints might be organized.

We should first observe that where the object-oriented architectural forms are concerned, the *Unified Modeling Language* (UML) is widely supported by a range of tools and well-defined forms [Rumbaugh, Jacobson & Booch 1999]. However, the term "unified" as used here refers to the drawing together of the ideas belonging to the three methodology "gurus" who wrote the standard, and there has subsequently been some significant empirically based questioning about its complexity and general usefulness [Moody 2009, Budgen et al. 2011]. Hence, given its wide recognition, we have mainly illustrated the viewpoints with forms that may be less familiar to those accustomed to the UML, in order to demonstrate the breadth of the concepts.

3.2.1 The constructional viewpoint. Various forms of "object" and "class" diagram have been developed, although most tend to be broadly similar in form to the UML *class diagram*. Indeed, in many ways, they closely resemble the Entity-Relationship Diagrams (ERDs) used for data modelling, although obviously the range of relationships included in such forms is more related to object and class interactions, such as aggregation, uses, and inheritance.

In contrast, Figure 5 shows an example of a *Structure Chart*, a form that is generally associated with a call-and-return form of architectural style. In this notation, the boxes represent subprogram units, and the lines joining them represent invocation (akin to the use of methods in object-oriented terminology).

Order and frequency of invocation are not shown, only their existence. The small arrows provide details of the parameters being passed—there are other drawing conventions used with this form, as well as some variations (some authors prefer to provide a table detailing the parameters). This form of "call graph" sometimes has another role,: it is often created by maintainers of software (either manually or via support tools) in order to gain a clearer visualization of the hierarchy of units within an existing system.

3.2.2 The functional viewpoint. As indicated in Figure 5, this aspect of a system is probably the most difficult one to capture in diagrammatical form. *Data Flow Diagrams* (DFDs) capture functionality in terms of how the operations of the system affect the forms of information it holds. (Earlier, Tom De Marco suggested that this form was much older than computing, and was certainly used in the early 1990s to model how teams of clerks processed things like insurance claims.)

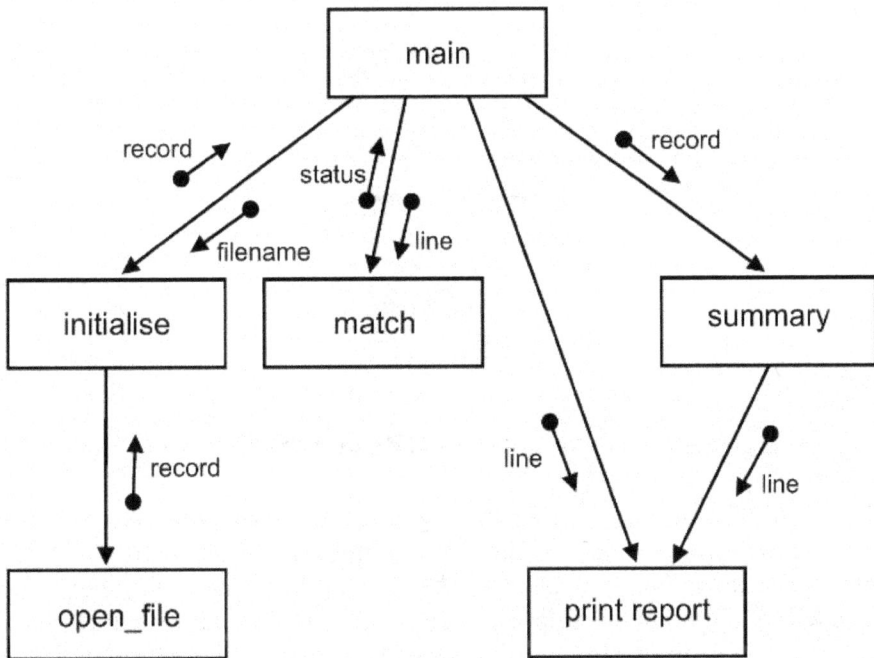

Figure 5: Example of a structure chart

A related but different way to describe function is in terms of workflows (focusing on the tasks rather than the data). Figure 6 shows one of the UML notations used for this (the *Activity Diagram*) being used to describe part of the process of conducting a systematic review. (*See [Budgen et al. 2011] for an example of this form of study.*) Here the focus is upon the activities being performed by the researchers and the related flows are divided and then recombined.

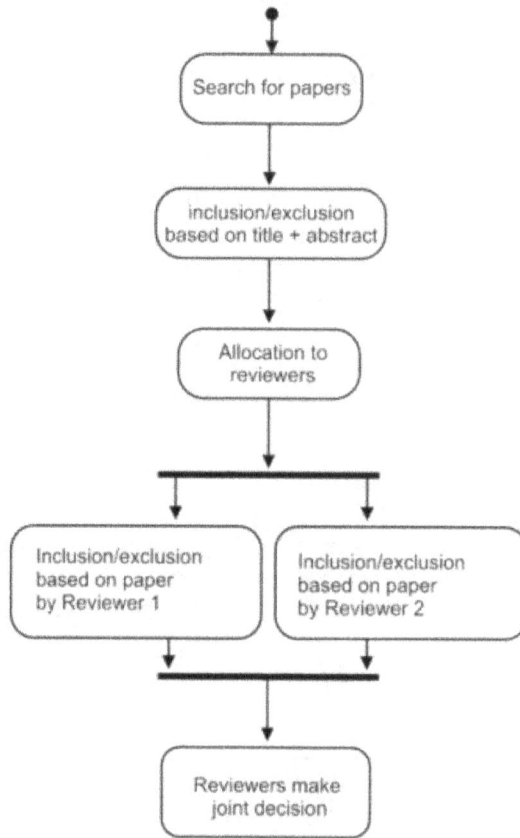

Figure 6: Example of a UML activity diagram

While this form of model is often largely associated with the data-centered repository architectural style and the use of databases in particular, it does have wider applications since we may also be interested in modelling how the data entities within a system are related. Such models might be hierarchical (i.e., how the information is associated with some system-level entity mapped onto lower level structures).

Here states can be combined to form "super states" and that can be decomposed into sub states. Our figure labels only a few of the transitions in order to keep it reasonably clear—the reader might wish to complete the missing ones. Here the focus is upon the activities being performed by the researchers and the related flows are divided and then recombined.

3.2.3 The data-modelling viewpoint. An example of the classic *Entity-Relationship Diagram (ERD)* is shown in Figure 7.

There are different drawing conventions associated with ERDs and this one uses a fairly long-established set of conventions by which the entities are shown as boxes, attributes of entities are placed in boxes with rounded corners, and the relationship is shown as numbers on the lines linking to the entities. (Here, some *n* aircraft can be held in the stack.)

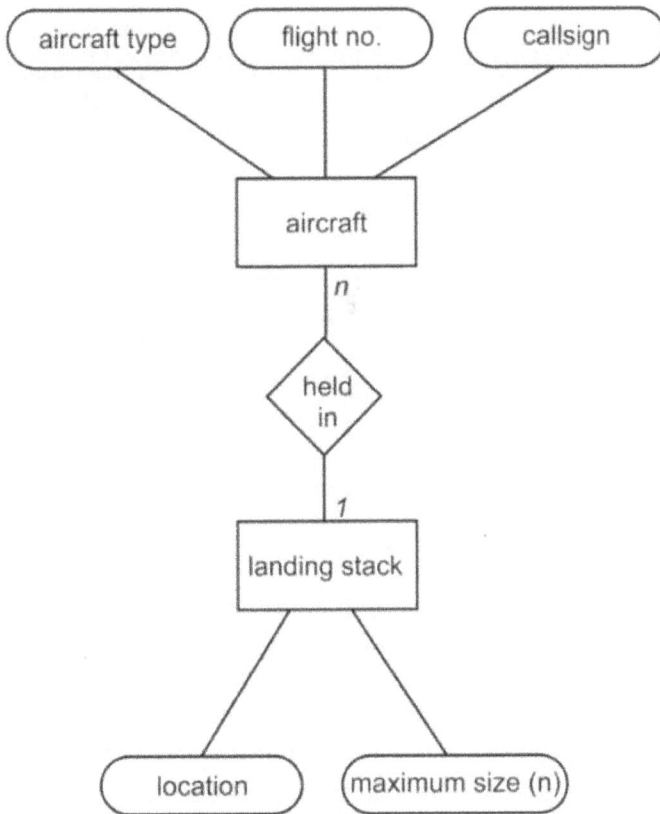

Figure 7: Example of an entity-relationship diagram

4. Organizing the Design Process

In Section 1.3, we discussed the concept of *knowledge schema*, with expert designers owning a richer (and more organized) set of such schema than less experienced designers. Since a knowledge schema is an internal representation of that knowledge, one of the challenges since the early days of software development has been to find ways of codifying that knowledge in such a way that the less experienced designer can learn design skills as quickly and effectively as possible. While the ideal might be that of a "design studio," where the "novice" can sit beside and learn from the "master," this is rarely a practical option. Indeed, expert designers are likely to be a rare commodity in most organizations [Curtis, Krasner and Iscoe 1988].

Figure 8 shows an example of a *statechart* [Harel 1987] being used to model the part of an air traffic control system concerned with "stacking" of aircraft that may not yet be able to land. We might also note that the form of the State Diagram that is used in the UML differs relatively little from this.

The earliest forms used for knowledge transfer were often termed "SED methods," using what we often term a *plan-driven* approach. As experience of design grew, and the range of software architectural styles expanded with technology, so did ideas about how design knowledge could usefully be organized.

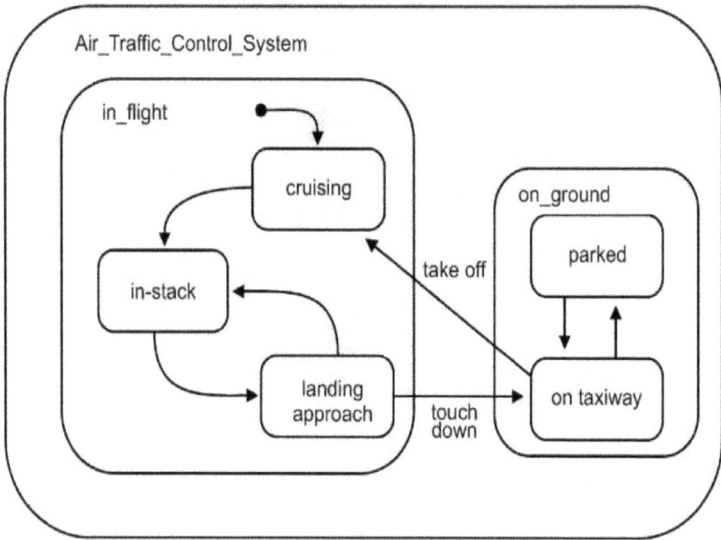

Figure 8: Example of a statechart

Design patterns offer an approach that is often considered more appropriate for object-oriented forms, and *component-based* design approaches have tried to employ a "black box" model that in some ways approximates the way that electronic hardware design is organized. The classical approach of the design method was often seen as being too closely linked to the waterfall model of development, and *Agile methods* have subsequently emerged as one means of making the SED process more responsive to its environment (including the business needs of the customer for the software).

In the following sections, we introduce each of these concepts and identify some key examples and references. Because space precludes an in-depth exposition of their features, the discussion necessarily has to be at a high level of abstraction. The final section discusses the role of design support tools and their limitations.

4.1 Plan-driven design.

Plan-driven design — essentially structure design knowledge as follows:

- A set of procedures that should be followed in order to create a "design model" that eventually evolves into the actual design plan.

- A set of descriptions, usually in the form of diagrams, that are used to represent the design model in various stages of evolution.

- A set of heuristics that are based upon the experience of using the method and might relate to such things as how to go about creating the initial model for a particular problem, or how to adapt the process for specific needs.

Therefore, these three elements essentially represent the knowledge schema as conveyed through a method. The procedures are organized around a strategy. Usually this is based upon either a "top-down" (decomposition) form or a "bottom-up" (compositional) form.

One of the earliest (and extremely successful) design methods was called by various names but was generally known as *structured analysis and design*. Here the initial model was based upon analyzing the data flow involved in the whole system (using some form of *data flow diagram*) and the procedures were concerned with transforming this into a call-and-return form of model based upon a main program and subprograms (usually represented as a *structure chart*).

Here the procedures were concerned with constructing the original model and then performing what were usually referred to as *transaction analysis* (identifying the different functions in the model) and *transform analysis* (changing the form of the model to map onto executable structures). The heuristics helped with identifying such concepts as the "central transform" for a particular transaction. A good example of a textbook describing this process is authored by Page-Jones [1988].

The original strategy was essentially one of functional decomposition, but perhaps reflecting the growing size of systems and greater experience resulted in more compositional forms, such as "event partitioning."

While the model could be and was extended, especially with the addition of real-time modelling features such as *state transition diagrams*, it was essentially limited by the use of rather one-dimensional models and by being tied to an architectural style (call-and-return) that was gradually replaced by the now dominant object-oriented forms. Hence, evolution essentially ended in the late 1980s, although its basic influence should not be under-rated as has been noted by Avison and Fitzgerald [1995].

The emergence of the object-oriented paradigm created problems for plan-driven forms. While these had proved quite effective when designing around such architectural styles as call-and-return and distributed processes, objects represent a much more complicated end model.

We will side step the (sometimes-thorny) issue of exactly what constitutes an "object." The terminology relating to objects is now fairly well established, and for this discussion, we will assume that objects are created from classes and that objects provide for encapsulation of their internal state and have public methods that can be used to inspect or modify that state. Objects can also inherit part of their structure from parent objects.

Two characteristics from the above, rather brief, outlines have provided a substantial problem for plan-driven approaches. One is encapsulation, while the other is inheritance. Neither of these fits well into the forms of description or procedure that were used for earlier design methods, nor have both continued to provide particular challenges for methodologists.

Through the 1980s and 1990s, a wide variety of objected-oriented methods was developed. Those of the "first generation" were largely evolutionary in nature, in the sense that they derived many of their ideas from earlier forms and often attempted to use non-object-oriented forms of system analysis. Later methods were more revolutionary, in the sense of using quite different (and more complex) procedures than those of earlier methods, with a stronger emphasis upon composition.

A key problem, regardless of strategy, has been to identify the "right" objects to use for a given design problem. While this tends to favor a compositional strategy, determining the choice of objects is still a complex one. Indeed, Francis Détienne [2002] has noted, "early books on objected-oriented design emphasized how easy it was to identify objects, while later ones, often by the same authors, emphasize the difficulty of identifying them."

A very comprehensive review of this theme, including descriptions of some objected-oriented methods, is provided in Wieringa [1998]. Wieringa particularly noted that the use of forms such as DFDs was incompatible with object-oriented structuring because of the enforced separation of data storage and data processing. DFD did not map onto the encapsulation of data and related operations embodied in the objected-oriented model. He did also note that there was "overwhelming agreement that the decomposition must be represented by a class diagram, component behavior by a statechart, and component communications by sequence or collaboration diagrams," although the detailed forms of these varied quite extensively.

By the late 1990s, the ideas of the major players in the objected-oriented method domain (Booch, Jacobson, and Rumbaugh) converged to create the *Unified Process*, which can perhaps be considered as a "third generation" method. Their forms are much more complex—(*see Jacobson, Booch and Rumbaugh [1999]*)—and indeed, are almost an intermediate form between that of earlier plan-driven forms and the Agile methods that we discuss below. Certainly, this seems to represent the "outer limit" as far as the development of plan-driven methods is concerned.

4.2 Design patterns.

Design patterns offer a quite different way of codifying design experience for re-use by others. Unlike architectural patterns, which describe the overarching form of the whole system, a design pattern is usually concerned with organizing a design component. Although the idea of the SED pattern resonates with the idea of "labels for plans" identified in such empirical studies of designers as that of Adelson and Soloway [1985], the design pattern community has instead drawn its inspiration from the ideas of an architect, Christopher Alexander. [Alexander et al. 1977] characterize a pattern as:

- Describing a recurring problem.

- Describing the core of a solution to that problem.

- The solution capable of being reused many times without actually using it in exactly the same way twice.

For SED, the pioneering work that established the concept widely as well as providing a standard for cataloguing patterns is the book by Gamma et al. [1995], with the authors (and the book) often referred to as the "gang of four" or *GoF*. Their book catalogues some twenty-three design patterns. A recent survey of software developers with extensive experience of pattern use [Zhang and Budgen 2010] suggests that not all have proven to be equally useful (and six of them to be of very questionable use). There is little question that patterns such as *Observer* and *Abstract Factory* provide useful guidelines about how to structure systems for ease of extension and change.

The key point about a pattern is that it is not simply a template for plugging in a choice of objects. A pattern is a way of organizing part of a design and as such, needs to be realized in a manner that fits local requirements, whatever their form might be.

As defined by the GoF, patterns fall into two categories in terms of their *scope*: classes or objects, with most patterns addressing the use of objects. They are also categorized by their *purpose*, whereby:

- *Creational* patterns are concerned with how and when objects need to be created for some purpose.

- *Structural* patterns are concerned with the ways that objects and classes are composed together.

- *Behavioral* patterns address the interaction between objects/classes and the responsibility is shared between them.

While patterns are unquestionably a valuable addition to the designer's repertoire, the enthusiasm of the pattern community for finding and documenting new patterns needs to be regarded with some caution. In particular:

- Over-enthusiastic use by inexperienced designers may lead to poorly structured designs; Sommerville [2007] argues firmly that patterns are best employed by more experienced designers who are better able to recognize that a design situation truly is structured using a generic form.

- The impact of using patterns is apt to be found during maintenance activities. Evidence here is patchy, but the paper by Wendorff [2001] provides some illustrations of the hazards of misuse, taken directly from experience.

To illustrate the concept, we will briefly examine the example of the widely used *Observer (293)* pattern. (By convention, when referring to patterns from the GoF, the page number is often appended to its name.) Observer provides an example of an object behavioral pattern, which concerns objects rather than classes and addresses a problem that is related to the dynamic behavior of those objects.

The situation that it addresses is one where a change of state that occurs in one object requires that (a variable number of) other objects then need to be notified so that, where appropriate, they can change their state to reflect this. A good illustration of such a situation occurs with a spreadsheet (the "subject") and a data-graphing tool (the "observer") that is providing a chart of the data in the spreadsheet. If we change the data values in some of the spreadsheet cells, then we expect that the graph will change in response, and that it will do so without doing anything for us.

Figure 9 shows a simple object model that represents this one-to-many situation, usually referred to as a "publish-subscribe" model.

In essence, observers register with the relevant subject when a state change occurs in the subject. It issues a *notify ( )* message to all registered observers and the observers then perform appropriate *update ( )* operations to obtain the change that might affect them.

Patterns emphasise the use of composition and interfaces over inheritance. When identifying a pattern, the goal is therefore to identify the parts of a design that are likely to vary, and to encapsulate these so that these parts of the system can vary without affecting others (Freeman et al., 2004). This delegation of changeable elements then creates the required flexibility for patterns such as Observer. Observer also demonstrates the idea of loose coupling that is common to many patterns. It also means that the key information remains under the control of a single object.

4.3 Agile methods.

For plan-driven design methods, the procedures involved nearly always assume that a complete requirements specification is available at the start of the design process. They are therefore implicitly based upon a traditional "waterfall" model of development. Since many such methods were developed—with "data processing" applications in mind—this is reasonable. However, as the computing environment has evolved from mainframes through personal computers to the Internet, expectations of software have changed and become more fluid.

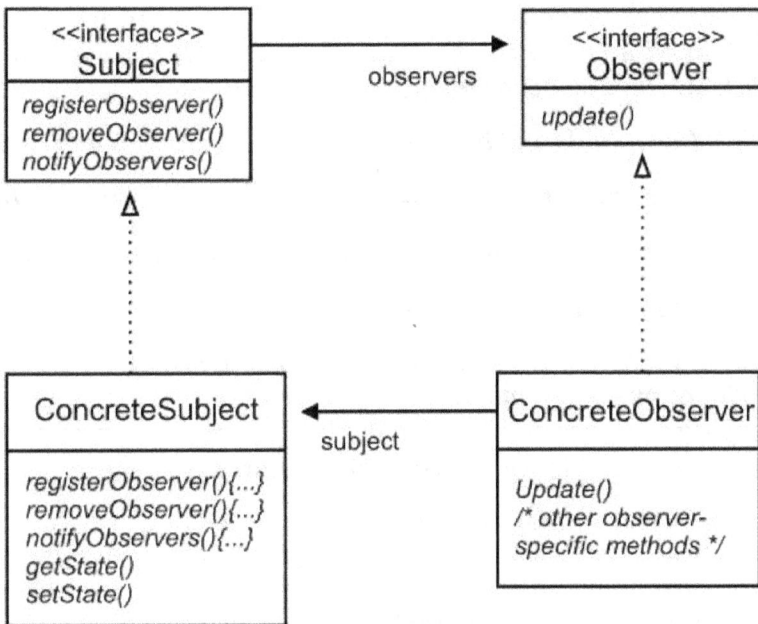

Figure 9: Structure of the observer pattern

The late 1980s and the 1990s saw the development of ideas about *rapid application development* (RAD), based at least in part upon the idea of developing a system through a series of incremental stages. Barry Boehm's *spiral model* of development [Boehm 1988] typifies this move from a waterfall-based context to one in which both design and implementation evolve in a stepwise fashion, adjusting as greater insight into the needs of the eventual end-users emerge.

The basis of the Agile movement was the idea that requirements were themselves "emergent" [Truex, Baskerville, and Klein 1999] and likely to be in a constant state of change as they adapted to continual changes in business needs.

With the new century, these ideas coalesced into more radical thinking (although still structured in this way) and led to the *Agile Manifesto* (*see Figure 10*).

A number of *Agile methods* have evolved from this, generally characterized by their use of incremental development models and close interaction with the end-user (customer). Probably the best-known forms are *Extreme Programming* (XP) developed by Kent Beck [2004], and *Scrum* [Schwaber and Beedle 2002] seems to be increasingly attracting attention. These two designers embody quite different approaches to addressing the ideas embodied in Figure 10.

We are uncovering better ways of developing software—first by doing it and later by helping others to do it. Through this work, we have come to value:

- *Individuals and interactions —over processes and tools.*

- *Working software —over comprehensive documentation.*

- *Customer collaboration —over contract negotiation.*

- *Responding to change —over following a plan.*

That is, while there is value in the items to the right, we value the items on the left more.

Figure 10: The Agile Manifesto

4.3.1 Extreme Programming (XP). XP is characterized by an emphasis upon its 12 basic practices rather than by a specific form of process model. Space precludes discussing all of these practices in detail, but below we briefly outline some of those that are probably better known (and which help to distinguish XP from other methods).

- *Test-first programming* — The XP practice is to write the tests before writing the code and then test continuously, at the end of each day, after each increment in the design.

- *Pair programming* — The best-known feature of XP. Code is written by two programmers working at a single machine, discussing their work as they go.

- *Collective ownership* — The code is owned by all of the team members, and they may make changes whenever deemed necessary.

- ***40-hour weeks*** — Iterations should be sized so that overtime is not needed, on the basis that tired programmers make mistakes.

While empirical studies of design methods are difficult to perform, some of the features of XP have been studied empirically, in particular pair programming. A secondary study (aggregating primary experimental studies) in the form of a meta-analysis was performed by Hannay et al. [2009]. The results of this were not strongly conclusive because of variations in the primary studies, but they did observe that—for problems that are more complex—the use of pair programming did seem to result in higher-quality software. However, for simpler problems, pair programming can be more time-consuming than solo programming.

4.3.2 Scrum. Unlike XP, *Scrum* is less concerned with technical issues and more with the management of the overall development process. A characteristic that Scrum shares with a number of RAD and Agile methods is the use of *time boxing*, by which emphasis is placed upon the use of development phases that use a fixed time interval, varying the resulting delivery of functionality as necessary. This is of course in contrast to plan-driven forms where functionality tends to be fixed and delivery times are varied as necessary.

Scrum projects are therefore organized around a series of fixed-term *sprints*, which are usually of 2-4 weeks in duration. Each sprint generates a new increment, and increments are grouped to create releases of the product. The list of development tasks is termed the *product backlog*, and a portion of the backlog is usually addressed in each sprint. The team is self-organizing and meets each day for a fifteen-minute *daily scrum* (the term comes from the huddle in the game of rugby football).

During the scrum, they review what has been accomplished since the previous meeting, identify what is to be done that day, and note any possible obstacles. Scheduled around this is a further set of formalized meetings that occur at the beginning and end of a sprint.

Scrum also distinguishes between different roles, which are characterized as being those of *pigs* or *chickens* (from the traditional model of the cooked breakfast for which the pig is committed and the chicken merely involved). In this context, roles defined as pigs are those that carry responsibility, and hence, chickens may not direct the activities of pigs.

4.4 Component-based development.

In other domains, the role of the *component* has been highly successful, implying well-defined and thoroughly tested functionality and interfaces, so enabling the designer to *reuse* such components in other systems.

In the 1990s, researchers began to explore how this concept might be employed for software. One problem that emerged was that of determining exactly what should be the key properties of a software component in order to enable the degree of reuse achieved in other domains. In Budgen [2003], one of the chapters examines this question and discusses the evolution of the component concept—a process that has continued. Some component-based software ideas are:

- Provision for *reuse* (implying a clear definition of interfaces needed to enable a "plug and play" role in which a component could simply be viewed as a "black box").

- Independence of delivery (a component should not have any "awareness" of its context).

- Existence of a component model that incorporates specific component interactions and composition standards.

- A composition standard that provides the necessary definitions of how components can be assembled to form larger structures.

Elements of these are examined in the discussions of components provided in papers such as [Brown and Short 1997], and books such as [Szyperski 1998] and [Heineman and Councill 2001].

However, while at one point it began to look as though a component "market" was emerging, this has not really developed as far as many expected or hoped. The reasons for this probably include at least the following:

- The commercial potential of a component market was probably undermined by the emergence of open source systems (and components), so that vendors of commercial components were disinclined to invest too extensively.

- The lack of adequate standards, in the sense that beyond a few specialized areas such as Java APIs, there was no agreed framework that component vendors and users could depend upon.

- The emergence of software service models and related concepts of service-oriented architecture (SOA). (To be addressed next.)

We are uncovering better ways of developing software by designing programs and helping others to perfect their development. Through this work, we have come to value the Agile Manifesto (*see Figure 10*).

Service models (and particularly web services) began to develop in the early 2000s. Their use of standards for interaction (such as the use of SOAP—the *simple object access protocol*), simplicity of interface, and platform independence offered an (admittedly constrained) approach to component assembly. Architecturally, service models provide a constrained form of distributed processes. However, because the computing work is undertaken by the service provider, they more readily lend themselves to creating commercial opportunities than components executed by the user on their own computer (Krafzig, Banke and Slama 2004). Emerging ideas about *cloud computing* take this aspect yet further.

As so often though, the service model, while appearing to overcome some of the limitations of preceding technologies, brings technical challenges of its own. Its distributed nature means that the end-user is dependent upon others to provide computing resources; indexing and locating services offers something of a semantic challenge; and there is a multiplicity of standards and many differing definitions of SOA. Software service models do certainly offer potential and they change the computing paradigm—but they have yet to demonstrate their effectiveness potential.

4.5 Design support tools.

Software tools can provide support for creative activities in many domains. The word processor provides features that are useful to the author; musical composition can be made easier by score-writing software that helps keep track of multiple parts; engineering has long had CAD (computer assisted design) software to help remove the more tedious drafting tasks.

Curiously, the one domain where design support tools have made little real progress is that of software engineering. There are a number of possible reasons for this: one is that the design of software involves manipulating abstractions, and in the early stages at least, this usually involved relatively little attention to detailed syntax and semantics. However, the examples of successful design support identified above are all taken from domains where design tends to take place using well-defined forms.

A second reason is the invisibility of our media. As observed above, our notations are essentially artificial and lack any well-defined or easily envisaged connections with the product itself. Related to this is the need for well-established descriptive standards. Musical notation, text, and 3-D engineering and drawing descriptions are all forms that have well-established conventions and standards. While for object-oriented design and for some other forms the UML has at least partially met this need, as we noted earlier, in its present form at least, this may just be one step on the path towards this objective.

Software design tools have tended to provide the means of drawing diagrams using such forms as the UML. However, it is still common to find that their design makes it difficult to create diagrams with non-standard syntax or forms. Yet, as we have also observed, SEDs rarely emerge "fully fledged" and expressed in a well-defined syntax. Indeed, although the challenge of developing such tools has long been recognized [Guindon and Curtis 1988, Reeves, Marashi, and Budgen 1995], progress with addressing this challenge seems to have been largely limited to developing tools for use in education (for example, *see [Dranidis 2007]*).

5. Discussion

Software design is a large and complex topic and an overview paper such as this can only provide an outline description of some of the key issues and developments included in the topic, together with some pointers to where the reader can obtain more detail.

The SEDer's repertoires of conceptual tools are both (1) quite extensive and (2) need care in its use. As the discussion of knowledge schema indicates, each designer has their own set of models, based on their own experiences, and experiences obtained from others—through whatever means are most appropriate. The sheer difficulty of undertaking empirical studies in this area has tended to limit our understanding of the effectiveness of our conceptual tools, although this situation is slowly changing. However, regardless of this, it is always important for the reader to be aware that SED does not lend itself to "silver bullets"—and indeed, conceptual tools are just that: they are aids that assist the designer in performing their own creative task, not a source of solutions in themselves.

Acknowledgements

A review paper such as this draws upon many sources and past discussions with colleagues and collaborators and my thanks to all of them for their help with exploring this complex and fascinating topic.

REFERENCES

- **[Adelson & Soloway 1985]** B. Adelson and E. Soloway, "The Role of Domain Experience in Software Design," *IEEE Trans. Software Eng.*, 11(11), 1985, pp. 1351-1360.

- **[Alexander et al. 1977]** C. Alexander, S. Ishikawa, M. Silverstein, M. Jacobson, I. Fiksdahl-King and S. Angel, *A Pattern Language*, Oxford University Press, Oxford, England, 1977.

- **[Avison & Fitzgerald 1995]** D.E. Avison and G. Fitzgerald, *Information Systems Development Methodologies, Techniques and Tools*, 2nd edition, McGraw-Hill, New York, 1995.

- **[Beck 2004]** Kent Beck, *Extreme Programming Explained: Embrace Change*, 2nd edition, Addison-Wesley, Reading, MA, 2004.

- **[Boehm 1988]** B.W. Boehm, "A spiral model of software development and enhancement," *IEEE Computer*, 21(5), 1988, pp. 61–72.

- **[Brooks 1987]** F.P. Brooks, Jr. "No Silver Bullet: Essence and Accidents of Software Engineering," *IEEE Computer*, April 1987, pp. 10-19.

- **[Brown & Short 1997]** A.W. Brown and K. Short, "On components and objects: The foundations of component-based development," *Proceedings of the 5th International Symposium on Assessment of Software Tools and Technologies*, IEEE Computer Society Press, 1997.

- **[Budgen 2003]** D. Budgen, *Software Design*, 2nd ed., Addison-Wesley, Reading, MA, 2003.

- **[Budgen et al. 2011]** D. Budgen, A.J. Burn, O.P. Brereton, B.A. Kitchenham, and R. Pretorius, "Empirical Evidence about the UML: A Systematic Literature Review," *Software: Practice & Experience*, 41(4), pages 363-392, 2011.

- **[Buschmann et al. 1996]** F. Buschmann, R. Meuniere, H. Rohnert, P. Summerland and M. Stall, *Pattern-Oriented Software Architecture: A System of Patterns*, Wiley, Hoboken, NJ, 1996.

- **[Curtis, Krasner, & Iscoe 1988]** B. Curtis, H. Krasner and N. Iscoe, "A field study of the software design process for large systems," *Comm. ACM*, 31(11), 1988, pp. 1268-1287.

- **[Détienne 2002]** F. Détienne, *Software Design—Cognitive Aspects*, Springer Practitioner Series, Heidelberg, Germany, 2002.

- **[Dranidis 2007]** D. Dranidis, "Evaluation of Student UML: An Educational Tool for Consistent Modelling with UML," *Proceedings of the Informatics Europe II Conference*, 2007, pp. 248-256.

- **[Fenton & Pfleeger 1997]** N.E. Fenton and S.L. Pfleeger, *Software Metrics: A Rigorous & Practical Approach*, 2nd ed., PWS Publishing Company, Boston, MA, 1997.

- **[Freeman, Sierra, & Bates 2004]** E. Freeman, K. Sierra, and B. Bates, *Head First Design Patterns*, O'Reilly Media, North Sebastopol, CA, 2004.

- **[Gamma et al. 1995]** E. Gamma, R. Helm, R. Johnson and J. Vlissides, *Design Patterns—Elements of Reusable Object-Oriented Software*, Addison-Wesley, Reading, MA, 1995.

- **[Garlan & Perry 1995]** D. Garlan and D.E. Perry, "Introduction to the special issue on software architecture," *IEEE Trans. on Software. Eng.*, 21(4), 1995, pp. 269-274.

- **[Guindon & Curtis 1988]** R. Guindon and B. Curtis, "Control of cognitive processes during software design: What tools are needed," *Proceedings of CHI'88*, ACM Press, 1988, pp. 263-268.

- **[Guindon 1990]** R. Guindon, "Knowledge exploited by experts during software system design," *Int. J. Man-Machine Studies*, (33), pp. 279-304.

- **[Hannay et al. 2009]** J.E. Hannay, T. Dybå, E. Arisholm and D.I.K. Sjøberg, "The effectiveness of pair programming: A meta-analysis," *Information & Software Technology*, (51), 2009, pp. 1110–1122.

- **[Harel 1987]** D. Harel, "Statecharts: a visual formalism for complex systems," *Science of Computer Programming*, 8, 1987, 231-274.

- **[Hayes-Roth & Hayes-Roth 1979]** R. Hayes-Roth and F. Hayes-Roth, "A Cognitive Model of Planning," *Cognitive Science*, (3), 1979, pp. 275-310.

- **[Heineman & Councill 2001]** G.T. Heineman and W.T. Councill, *Component-Based Software Engineering: Putting the Pieces Together*, Addison-Wesley, Reading, MA, 2001.

- **[Jacobson, Booch, & Rumbaugh 1999]** I. Jacobson, G. Booch, and J. Rumbaugh, *The Unified Software Development Process*, Addison-Wesley, Reading, MA, 1999.

- **[Krafzig, Banke, & Slama 2004]** D. Krafzig, K. Banke, and D. Slama, *Enterprise SOA: Service-Oriented Architecture Best Practices*, Prentice-Hall, Upper Saddle River, NJ, 2004.

- **[Kruchten 1994]** P.B. Kruchten, "The 4+1 view model of architecture," *IEEE Software*, 12(6), 1994, pp. 42-50.

- **[Moody 2009]** D.L. Moody, "The "'physics" of notations: Toward a scientific basis for constructing visual notations in software engineering," *IEEE Trans. on Software. Eng.*, 35(6), 2009, pp. 756-779.

- **[Page-Jones 1988]** M. Page-Jones, *The Practical Guide to Structured Systems Design*, 2nd edition, Prentice-Hall, Upper Saddle River, NJ, 1988.

- **[Parnas & Weiss 1987]** D.L. Parnas and D.M. Weiss, "Active Design Reviews: Principles and Practices," *J. Systems & Software*, (7), 1987, pp. 259-265.

- **[Reeves, Marashi, & Budgen 1995]** A.C. Reeves, M. Marashi, and D. Budgen, "A software design framework or how to support real designers," *Software Eng. Journal*, 10(4), 1995, pp. 141-155.

- **[Rittel & Webber 1984]** H.J. Rittel and M.M. Webber, "Planning Problems are Wicked Problems," N. Cross, eds., *Developments in Design Methodology*, Wiley, Hoboken, NJ, 1984, pp. 135-144.

- **[Rumbaugh, Jacobson, & Booch 1999]** J. Rumbaugh, I. Jacobson, and G. Booch, *The Unified Modeling Language Reference Manual*, Addison-Wesley, Reading, MA, 1999.

- **[Schwaber & Beedle 2002]** K. Schwaber and M. Beedle, *Agile software development with Scrum*, Prentice-Hall, Upper Saddle River, NJ, 2002.

- **[Shaw & Garlan 1996]** M. Shaw and D. Garlan, *Software Architecture: Perspectives on an Emerging Discipline*, Prentice-Hall, Upper Saddle River, NJ, 1996.

- **[Sommerville 2007]** I. Sommerville, *Software Engineering*, 8th edition, Addison-Wesley, Reading, MA, 2007.

- **[Szyperski 1998]** C. Szyperski, *Component Software: Beyond Object-Oriented Programming*, Addison-Wesley, Reading, MA, 1998.

- **[Taylor, Medvidovic, & Dashofy 2010]** R.N. Taylor, N. Medvidovic, and E.M. Dashofy, *Software Architecture: Foundations, Theory and Practice*, Wiley & Sons, Hoboken, NJ, 2010.

- **[Truex, Baskerville, & Klein 1999]** D. Truex, R. Baskerville, and H. Klein, "Growing systems in emergent organizations," *Comm. ACM*, 42(8), 1999, pp. 117–123.

- **[Wendorff 2001]** P. Wendorff, "Assessment of design patterns during software reengineering: Lessons learned from a large commercial project," *Proc. of Fifth Conference on Software Maintenance and Reengineering CSMR '01*, IEEE Computer Society Press, Los Alamitos, CA, 2001, pp. 77-84.

- **[Wieringa 1998]** R. Wiering, "A survey of structured and object-oriented software specification methods and techniques," *ACM Computing Surveys*, 30(4), 1998, pp. 459-527.

- **[Williams & Cockburn 2003]** L. Williams and A. Cockburn, "Agile software development: It's about feedback and change," *IEEE Computer*, 36(6), 2003, pp. 39–43.

- **[Zhang & Budgen 2010]** C. Zhang and D. Budgen, "What Do We Know about the Effectiveness of Software Design Patterns?" (A Survey of Experience about Design Patterns), *IEEE Transactions on Software Engineering*, 38(5), 2010, pp. 1213-1231.

Chapter 3

Model-Based Software Design for Concurrent and Real-Time Systems

Hassan Gomaa, PhD
George Mason University

Abstract

When designing concurrent and real-time systems, it is essential to blend object-oriented concepts with the concepts of concurrent processing. This paper describes a model-based software design method for designing concurrent and real-time systems, which integrates object-oriented and concurrent processing concepts and uses the UML notation.

Keywords: real-time systems, UML, concurrency, real-time software, design method, software product lines, software modeling.

1. Introduction

In model-based software design and development, software modeling is used as an essential part of the software development process. Models are built and analyzed prior to the implementation of the system, and are used to direct the subsequent implementation. A better understanding of a system can be obtained by considering the multiple views [Gomaa 2004, Gomaa 2006], such as requirements models, static models, and dynamic models of the system. A graphical modeling language such as UML helps in developing, understanding, and communicating the different views.

Real-time systems are reactive systems, so that control decisions are often state dependent, hence the importance of finite state machines in the design of these systems. Real-time systems typically need to process concurrent inputs from many sources, hence the importance of concurrent software design. They have real-time throughput and/or response time requirements, so there is a need to analyze the performance of real-time designs. Furthermore, there is a need to integrate real-time technology with modern software engineering concepts and methods.

This paper provides an overview of designing real-time embedded software systems. It starts by providing an overview of concurrent processing concepts in Section 2. In section 3, run-time support for concurrent and real-time systems is briefly discussed. Section 4 presents an overview of concurrent and real-time design methods. With this background, an overview of a model-based software design method for distributed and real-time embedded systems is given in Section 5. The COMET method [Gomaa 2000, Gomaa 2011] integrates object-oriented and concurrent processing concepts, and uses the Unified Modeling

Language (UML) notation. Section 6 describes software architectural patterns for real-time control. Section 7 describes the performance analysis of real-time software designs. Section 8 describes the design of real-time embedded software product lines.

2. Concurrent Processing Concepts

A characteristic of all real-time embedded systems is that of concurrent processing; that is, many activities occurring simultaneously, whereby frequently, the order of incoming events is not predictable. Consequently, as real-time embedded systems deal with several concurrent activities, it is highly desirable for a real-time embedded system to be structured into concurrent tasks (also known as concurrent processes or threads). This section describes the concepts of the concurrent task, and the communication and synchronization between co-operating tasks. For more information, refer to [Bacon 2003, Magee and Kramer 2006, Silberschatz and Galvin. 2008, and Tanenbaum 2008].

A concurrent task (also known as a concurrent process) represents the execution of a sequential program or sequential component of a concurrent program. A concurrent system consists of several tasks executing in parallel. Each task deals with one sequential thread of execution. Concurrency in a software system is obtained by having multiple asynchronous tasks, running at different speeds. From time to time, the tasks need to communicate and synchronize their operations with each other. The concurrent tasking concept has been applied extensively in the design of operating systems, real-time systems, interactive systems, distributed systems, parallel systems, and in simulation applications [Bacon 2003].

3. Run-Time Support for Concurrent Tasks

Runtime support for concurrent processing may be provided by:

- **Kernel of an operating system** — This has the functionality to provide services for concurrent processing. In some modern operating systems, a micro-kernel provides minimal functionality to support concurrent processing, with most services provided by system level tasks.

- **Run-time support system** — For a concurrent language.

- **Threads package** — Provides services for managing threads (lightweight processes) within heavyweight processes.

For more information, refer to [Gomaa 2000].

4. Survey of Design Methods for Concurrent and Real-Time Systems

For the design of concurrent and real-time systems, a major contribution came in the late seventies with the introduction of the MASCOT notation [Simpson et al. 1979] and later the MASCOT design method [Simpson 1986].

Based on a data flow approach, MASCOT formalized the way tasks communicate with each other, via either channels for message communication or pools (information hiding modules that encapsulate shared data structures).

The 1980s saw a general maturation of software design methods, and several system design methods were introduced. Parnas's work with the Naval Research Lab, in which he explored the use of information hiding in large-scale software design, led to the development of the Naval Research Lab (NRL) Software Cost Reduction Method [Parnas, Clements, and Weiss 1984]. Work on applying Structured Analysis and Structured Design to concurrent and real-time systems led to the development of Real-Time Structured Analysis and Design (RTSAD) [Ward 1985, Hatley 1988] and the *Design Approach for Real-Time Systems (DARTS)* [Gomaa 1984] methods.

Another software development method to emerge in the early 1980s was Jackson System Development (JSD) [Jackson 1983]. JSD was one of the first methods to advocate that the design should model reality first and, in this respect, predated the object-oriented analysis methods. The system is considered a simulation of the real world and is designed as a network of concurrent tasks, where each real-world entity is modeled by means of a concurrent task. JSD also defied the then-conventional thinking of top-down design by advocating a scenario-driven behavioral approach to software design. This approach was a precursor of object interaction modeling, an essential aspect of modern object-oriented development.

The early object-oriented analysis and design methods emphasized the structural aspects of software development through information hiding and inheritance but neglected the dynamic aspects, and hence were less useful for real-time design. A major contribution by the Object Modeling Technique [Rumbaugh et al. 1991] was to demonstrate clearly that dynamic modeling was equally important. In addition to introducing the static modeling notation for the object diagrams, OMT showed how dynamic modeling could be performed with statecharts (hierarchical state transition diagrams originally conceived by [Harel and Gary 1996], and [Harel and Politi 1998] for showing the state-dependent behavior of active objects and with sequence diagrams to show the sequence of interactions between objects.

The CODARTS (Concurrent Design Approach for Real-Time Systems) method [Gomaa 1993] built on the strengths of earlier concurrent design, real-time design, and early object-oriented design methods. These included Parnas's NRL Method, Booch's Object-Oriented Design [Booch 2007], JSD, and the DARTS method by emphasizing both information hiding module structuring and task structuring. In CODARTS, concurrency and timing issues are considered during task design and information hiding issues are considered during module design.

Octopus [Awad, Kuusela, and Ziegler 1996] is a real-time design method based on use cases, static modeling, object interactions, and statecharts. By combining concepts from Jacobson's use cases with Rumbaugh's static modeling and statecharts, Octopus anticipated the merging of the notations that is now the UML. For real-time design, Octopus places particular emphasis on interfacing to external devices and on concurrent task structuring.

ROOM – Real-Time Object-Oriented Modeling – [Selic, Gullekson, and Ward 1994] is a real-time design method that is closely tied in with a CASE (Computer Assisted Software Engineering) tool called ObjecTime. ROOM is based around actors, which are active objects that are modeled using a variation on statecharts called ROOM charts. A ROOM model, which has been specified in sufficient detail, may be executed. Thus, a ROOM model is operational and may be used as an early prototype of the system.

Buhr [1996] introduced an interesting concept called the use case map (based on the use case concept) to address the issue of dynamic modeling of large-scale systems. Use case maps consider the sequence of interactions between objects (or aggregate objects in the form of subsystems) at a larger grained level of detail than do communication diagrams.

For UML-based real-time software development, Douglass [1999, 2004] has provided a comprehensive description of how UML can be applied to real-time systems. The 2004 book describes applying the UML notation to the development of real-time systems. The 1999 book is a detailed compendium covering a wide range of topics in real-time system development, including safety-critical systems, interaction with real-time operating systems, real-time scheduling, behavioral patterns, and real-time debugging and testing.

5. A Model-Based Software Design Method for Concurrent and Real-Time Embedded Systems

5.1 Introduction.

Most books on object-oriented analysis and design only address the design of sequential systems or omit the important design issues that need to be addressed when designing real-time and distributed applications [Gomaa 2000, Bacon 2003]. It is essential to blend object-oriented concepts with the concepts of concurrent processing in order to successfully design these applications. This article describes some of the key aspects of the use of COMET to successfully design model-based software design methods for real-time embedded and distributed systems.

COMET integrates object-oriented and concurrent processing concepts and uses the Unified Modeling Language (UML) notation [Rumbaugh et al. 2005]. It also describes the decisions made explaining how to use the UML notation to address the design of concurrent, distributed, and real-time embedded systems. Examples are given from a Pump Monitoring and Control System, which is depicted using the UML 2 notation.

5.2 The COMET method.

COMET is a Concurrent Object Modeling and Architectural Design Method for the development of concurrent applications, in particular distributed and real-time embedded applications [Gomaa 2000, Gomaa 2011]. As the UML is now the standardized notation for describing object-oriented models [Booch et al. 2005, Rumbaugh et Al. 2005, Jacobson et al. 2000], the COMET method uses the UML notation throughout.

The COMET Object-Oriented Software Life Cycle is highly iterative. In the Requirements Modeling phase, a use case model is developed in which the functional requirement of the system is defined in terms of actors and use cases.

In the Analysis Modeling phase, static and dynamic models of the system are developed. The static model defines the structural relationships among problem domain classes. Object structuring criteria are used to determine the objects to be considered for the analysis model. A dynamic model is then developed in which the use cases from the requirements model are refined to show the objects that participate in each use case and how they interact with each other. In the dynamic model, state dependent objects are defined using statecharts.

In the Design Modeling phase, an Architectural Design Model is developed. Subsystem structuring criteria are provided to design the overall software architecture. For distributed applications, a component-based development approach is taken, in which each subsystem is designed as a distributed self-contained component.

The emphasis is on the division of responsibility between clients and servers, including issues concerning the centralization vs. distribution of data and control, and the design of message communication interfaces, including synchronous, asynchronous, brokered, and group communication. Each concurrent subsystem is then designed, in terms of active objects (tasks) and passive objects. Task communication and synchronization interfaces are defined. The performance of real-time designs is estimated using an approach based on rate monotonic analysis [SEI 1993].

Distinguishing features of the COMET method emphasize:

- Structuring criteria to assist the designer at different stages of the analysis and design process: subsystems, objects, and concurrent tasks.

- Dynamic modeling, both object communication diagrams and statecharts, describe in detail how object communication diagrams and statecharts relate to each other.

- Distributed application design, which addresses the design of configurable distributed components and inter-component message communication interfaces.

- Concurrent design, addressing in detail task structuring and the design of task interfaces.

- Performance analysis of real-time designs using real-time scheduling.

COMET emphasizes the use of structuring criteria at different stages in the analysis and design process. Object structuring criteria are used to help determine the objects in the system, subsystem structuring criteria are used to help determine the subsystems, and concurrent task structuring is used to help determine the tasks (active objects) in the system. UML stereotypes are used throughout to clearly show the use of the structuring criteria.

The UML Notation supports requirements, analysis, and design concepts. The COMET method separates requirements, analysis, and design activities. Requirements modeling defines the functional requirements of the system. COMET differentiates analysis from design as follows: analysis is breaking down or decomposing the problem so that it is understood better; design is synthesizing or composing (putting together) the solution. These activities are now described in more detail.

5.3 Requirements modeling with UML.

In the requirements model, the system is considered to be a black box. The Use Case Model is developed in which the functional requirements of the system are defined in terms of use cases and actors. This section describes the use of actors in real-time applications.

There are several variations on how actors are modeled [Jacobson 1992, Booch 2007, Fowler 2004, Gomaa 2011]. An actor is very often a human user. In many information systems, humans are the only actors. It is also possible in information systems for an actor to be an external system. In real-time and distributed applications, an actor can also be an external I/O device or a timer. External I/O devices and timer actors are particularly prevalent in real-time embedded systems, where the system interacts with the external environment through sensors and actuators.

A human actor may use various I/O devices to interact physically with the system. In such cases, the human is the actor and the I/O devices are not actors. In some cases, however, it is possible for an actor to be an I/O device. This can happen when a use case does not involve a human, as often happens in real-time applications.

An actor can also be a timer that periodically sends timer events to the system. Periodic use cases are needed when certain information needs to be output by the system on a regular basis. This is particularly important in real-time systems, although it can also be useful in information systems. Although some methodologists consider timers internal to the system, it is more useful in real-time application design to consider timers as logically external to the system and to treat them as primary actors that initiate actions in the system.

An example of a use case model from the Pump Monitoring and Control System is given in Figure 1, in which there are two use cases, Control Pump and View

Pump Status. There are five actors, three representing the three external sensors, one clock actor, and an external user actor, the Operator.

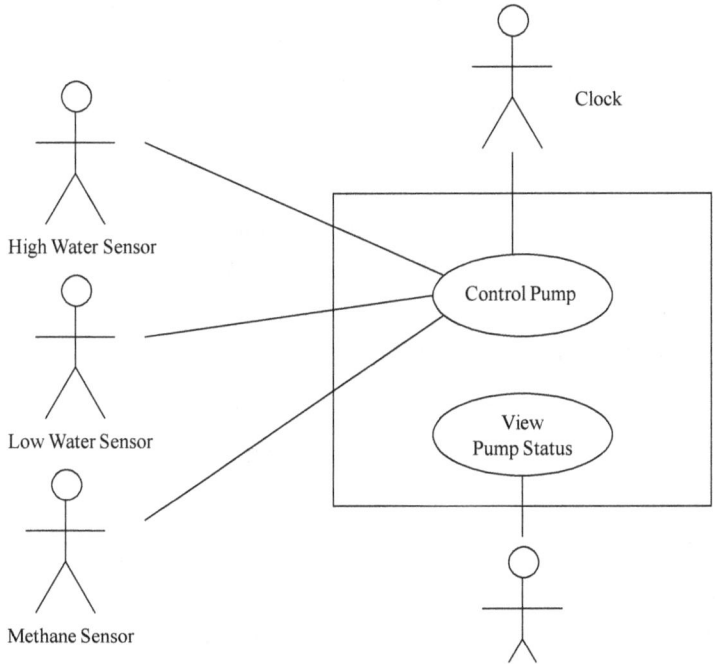

Figure 1: Use case model for pump monitoring and control system

5.4 Analysis modeling with UML.

This section describes some of the interesting aspects of COMET for analysis modeling. In particular, this section describes static modeling of the system context, stereotypes to represent object structuring decisions made by the analyst, and consistency checking between multiple views of a dynamic model.

5.4.1 Static modeling. For real-time applications, it is particularly important to understand the interface between the system and the external environment, which is referred to as the *system context*. In structured analysis [Yourdon 1989], the system context is shown on a *system context diagram*. The UML notation does not explicitly support a system context diagram. However, the system context may be depicted using either a static model or a communication model [Douglass 1999]. A *system context class diagram* provides a more detailed view of the system boundary than a use case diagram.

Using the UML notation for the static model, the system context is depicted showing the system as an aggregate class with the stereotype «software system», and the external environment is depicted as external classes to which the system has to interface. External classes are categorized using stereotypes. (*See description in Section 5.4.2.*) An external class can be an «external input device», an «external output device», an «external I/O device», an «external user», an «external system», or an «external timer». For a real-time system, it is desirable to identify low-level external classes that correspond to the physical I/O devices to which the system has to interface. These external classes are depicted with the stereotype «external I/O device».

An example of a system context class diagram from the Pump Monitoring and Control System is given in Figure 2. There are three external input device classes, namely the three sensors, one external output device class, the pump engine, one external timer class, and one external user class.

During the analysis modeling phase, static modeling is also used for modeling data-intensive classes [Rumbaugh 1991].

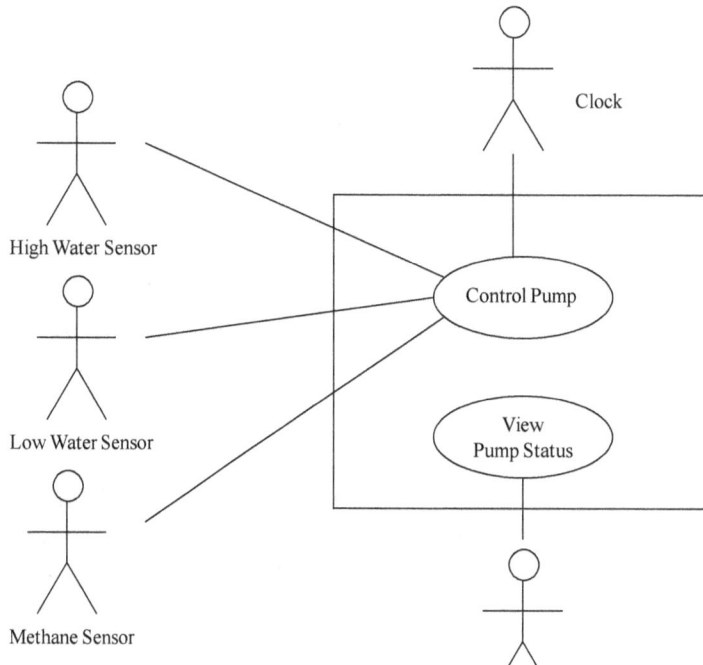

Figure 2: Use case model for pump monitoring and control system

5.4.2 Object structuring. Object structuring criteria are provided to assist the designer in structuring a system into objects. Several object-based and object-oriented analysis methods provide criteria for determining objects in the problem domain [Booch 1994, Coad 1991, Gomaa 1993, Jacobson 1992, Parnas 1984, Shlaer and Mellor 1988]. The COMET object structuring criteria build on these methods.

In object structuring, the goal is to categorize objects in order to group together objects with similar characteristics. Whereas classification based on inheritance is an objective of object-oriented modeling, it is essentially tactical in nature. Categorization, however, is a strategic classification. A decision is made to organize classes into certain groups because most software systems have these kinds of classes and categorizing classes in this way helps us understand the system we are to develop.

UML stereotypes are used to distinguish among the different kinds of application classes. A *stereotype* is a subclass of an existing modeling element, in this case an application class, which is used to represent a usage distinction, in this case the kind of class. A stereotype is depicted in guillemets, e.g., «control». An instance of a stereotype class is a stereotype object, which can also be shown in guillemets. Thus, an application class can be categorized as an «entity» class, which is a persistent class that stores data, a «boundary» class, which interfaces to and communicates with the external environment, and a «control» class. In addition, the applicant class can provide the overall coordination for the objects that participate in a use case, or an «application logic» class, which encapsulates algorithms separately from the data being manipulated.

Real-time systems will have many device interface classes to interface to the various sensors and actuators. They will also have complex state-dependent control classes because these systems are highly state dependent.

5.4.3 Dynamic modeling. For concurrent, distributed, and real-time applications, dynamic modeling is of particular importance. UML does not emphasize consistency checking between multiple views of the various models. Nevertheless, during dynamic modeling, it is important to understand how the finite state machine model, depicted using a statechart [Harel 1988, Harel et al. 1996, Harel et al. 1998] that is executed by a state-dependent control object, relates to the interaction model, which depicts the interaction of this object with other objects.

State Dependent Dynamic Analysis addresses the interaction among objects that participate in state-dependent use cases. A state-dependent use case has a state-dependent control object, which executes a statechart, providing the overall control and sequencing of the use case. The interaction is used among the objects that participate in the use case on a communication diagram or sequence diagram.

The statechart needs to be considered in conjunction with the communication diagram. In particular, it is necessary to consider the messages that are received and sent by the control object, which executes the statechart. An input event into the control object on the communication diagram must be consistent with the same event depicted on the statechart. The output event (which causes an action, enables or disables activity) on the statechart must be consistent with the output event shown on the communication diagram.

An example of the communication diagram for the Control Pump use case is given in Figure 3 and of the statechart for the Pump Control object is shown in Figure 4. In Figure 3, there are two input objects, High Water Sensor Interface and Low Water Sensor Interface, which receive inputs from the external input devices. There is one output object, Pump Engine Interface, which outputs to the external output device. There is one state-dependent control object, Pump Control, which executes the statechart in Figure 4. Finally, there is one timer object. Message inputs to the Pump Control object, such as High Water Detected, in Figure 3 is an example of an event that causes state changes on the statechart in Figure 4. Actions in Figure 4, such as Start Pump and Stop Pump, correspond to output messages from the Pump Control object in Figure 4.

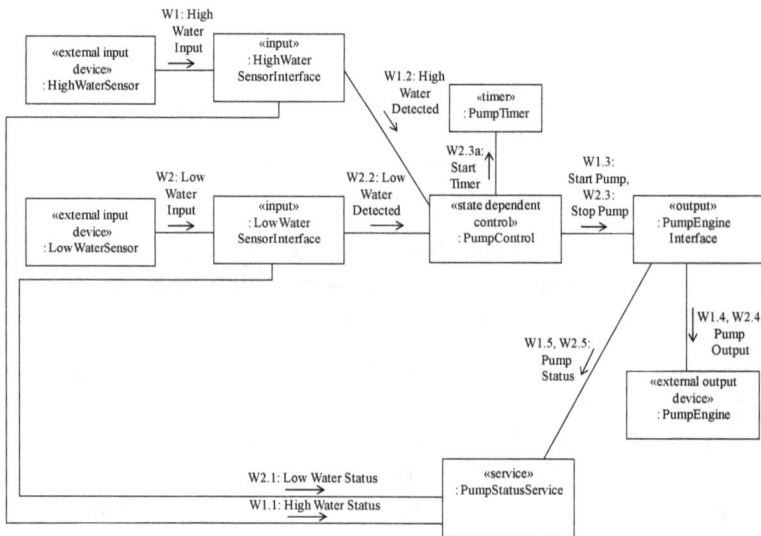

Figure 3: Communication diagram for control-pump use case

5.5 Design modeling.

This section describes some of the interesting aspects of COMET for design modeling. In particular, this section describes the consolidation of communication diagrams to synthesize an initial software design, subsystem structuring using packages, distributed application design, concurrent task design, and the design of connectors using monitors.

5.5.1 The transition from analysis to design. In order to transition from analysis to design, it is necessary to synthesize an initial software design from the analysis carried out so far. In the analysis model, a communication diagram is developed for each use case. The *integrated communication diagram* is a synthesis of all the communication diagrams developed to support the use cases. The consolidation performed at this stage is analogous to the robustness analysis performed in other methods [Jacobson 1992, Rosenberg 1999]. These other methods use the static model for robustness analysis, whereas COMET emphasizes the dynamic model, as this addresses the message communication interfaces, which is crucial in the design of real-time and distributed applications.

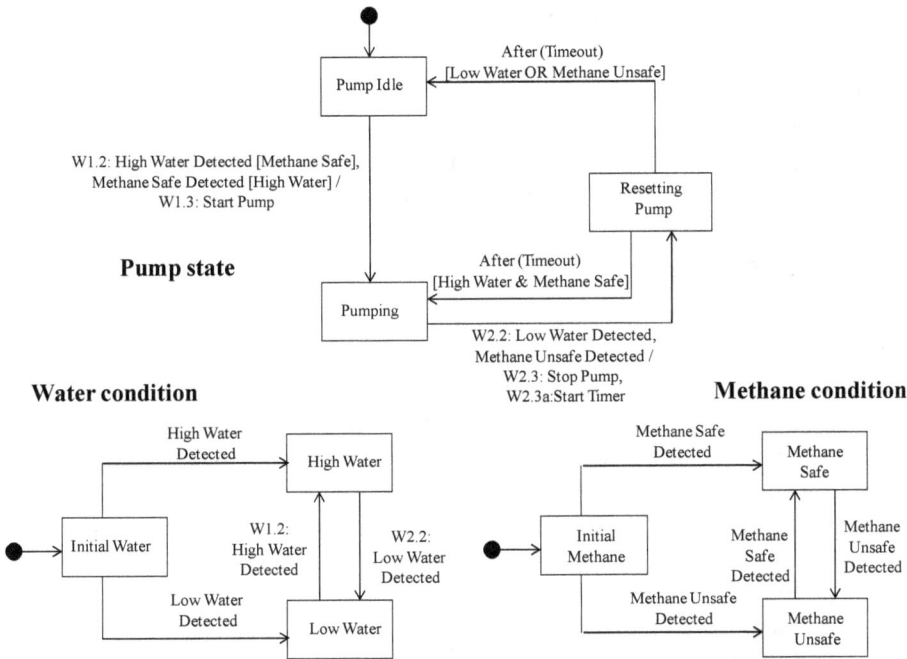

Figure 4: Pump control statechart

5.5.2 Software architectural design. During Software Architectural Design, the system is decomposed into subsystems and the interfaces between the subsystems are defined [Shaw 1996, Taylor 2009]. A system is structured into subsystems, which contain objects that are functionally dependent on each other. The goal is to have objects with high coupling among each other in the same subsystem, while objects that are weakly coupled are in different subsystems. A subsystem can be considered a composite or aggregate object that contains the simple objects that compose that subsystem.

5.5.3 Concurrent communication diagrams. In the UML 2 notation, an active object or task is depicted as a box with two parallel lines on the left and right sides of the object box. An active object has its own thread of control and executes concurrently with other objects. This is in contrast to a passive object, which does not have a thread of control.

A passive object only executes when another object (active or passive) invokes one of its operations. In this paper, we refer to an active object as a task and a passive object as an object.

Tasks are depicted on *concurrent communication diagrams*, which depict the concurrency concerns of the system [Douglass 2004], [Gomaa 2011]. On a concurrent communication diagram, a task is depicted as a box with thick black lines while a passive object is depicted as a box with thin black lines. In addition, decisions are made about the type of message communication between tasks, asynchronous or synchronous, with, or without reply.

5.5.4 Architectural Design of Distributed Real-Time Systems. Distributed real-time systems execute on geographically distributed nodes supported by a local or wide area network. With COMET, a distributed real-time system is structured into distributed subsystems, where a subsystem is designed as a configurable component and corresponds to a logical node. A subsystem component is defined as a collection of concurrent tasks executing on one logical node. As component subsystems potentially reside on different nodes, all communication between component subsystems must be restricted to message communication.

Tasks in different subsystems may communicate with each other using several different types of message communication (Figure 5) including asynchronous communication, synchronous communication, client/server communication, group communication, brokered communication, and negotiated communication. The configuration of the distributed real-time system is depicted on a deployment diagram, as shown in Figure 6, which shows the three subsystems depicted as distributed nodes in a distributed configuration.

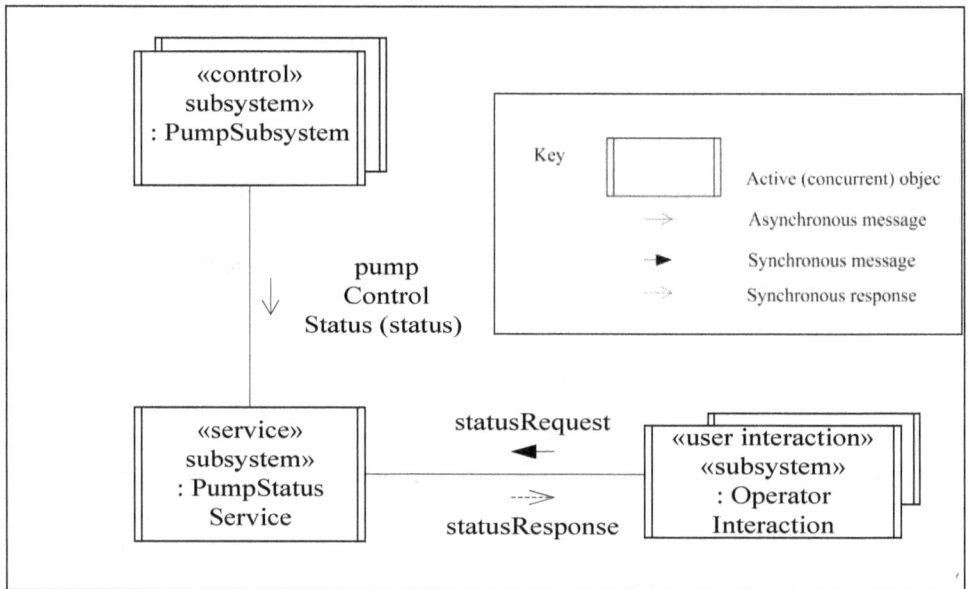

Figure 5: Distributed software architecture

5.5.5 Task structuring. During the task structuring phase, each subsystem is structured into concurrent tasks and the task interfaces are defined. Task structuring criteria are provided to assist in mapping an object-oriented analysis model of the system to a concurrent tasking architecture. Following the approach used for object structuring, stereotypes are used to depict the different kinds of tasks. Each task is depicted with two stereotypes; the first is the object role criterion, determined during object structuring as described in Section 5.4.2. The second stereotype is used to depict the type of concurrency.

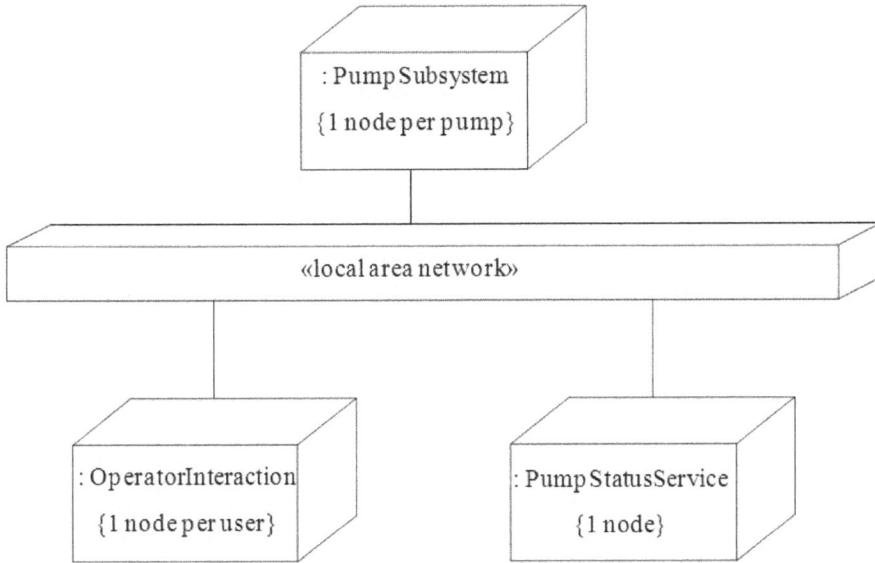

Figure 6: Distributed system configuration

During concurrent task structuring, if an object in the analysis model is determined to be active, it is categorized further to show its concurrent task characteristics. For example, an active «I/O» object is concurrent and is categorized further using a second stereotype as one of the following: an «event driven» task, a «periodic» task, or a «demand driven» task. Stereotypes are also used to depict the kinds of devices to which the concurrent tasks interface. Thus, an «external input device» is further classified, depending on its characteristics, into an «event driven» external input device or a «passive» external input device.

An event driven I/O task is needed when there is an event driven (also referred to as interrupt driven) I/O device to which the system has to interface. The event driven I/O task is activated by an interrupt from the event driven device.

While an event driven I/O task deals with an event driven I/O device, a periodic I/O interface task deals with a passive I/O device, where the device is polled on a regular basis. In this situation, the activation of the task is periodic but its function is I/O related. The periodic I/O task is activated by a timer event, performs an I/O operation, and then waits for the next timer event.

An example of task architecture for the Pump Monitoring and Control System is given in Figure 7. There are four tasks in the Pump Subsystem. There is one periodic input task, Methane Sensor Interface, a periodic temporal clustering task, Water Sensor, a demand driven control clustering task, Pump Controller, and a periodic timer task, Pump Timer.

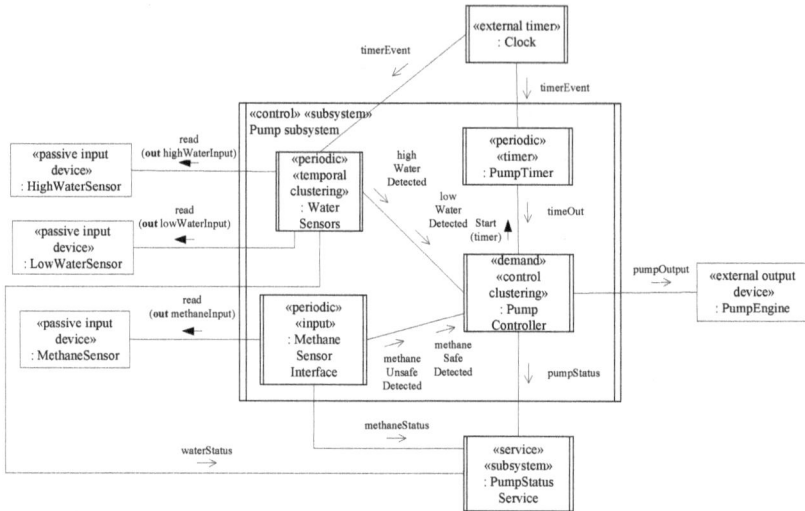

Figure 7: Pump system — task architecture

5.5.6 Detailed software design. In this step, the internals of composite tasks that contain nested objects are designed, detailed task synchronization issues are addressed, connector classes are designed that encapsulate the details of inter-task communication, and each task's internal event sequencing logic is defined. An example of the detailed design of a composite task is given in Figure 8.

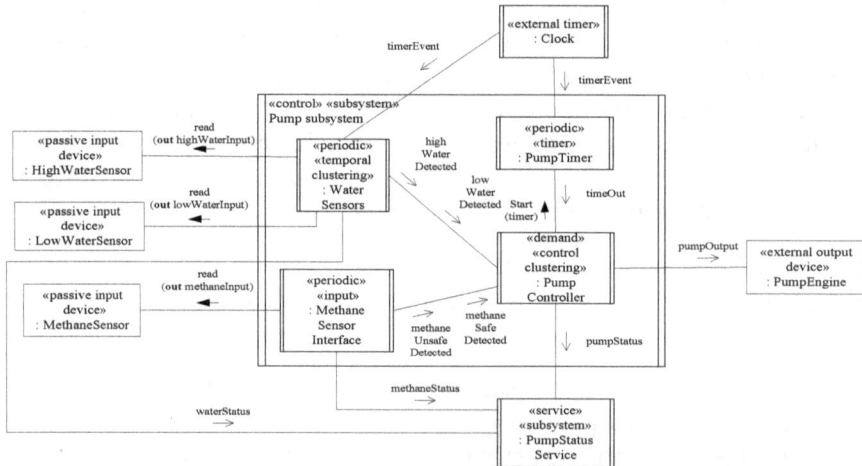

Figure 8: Water sensors — temporal clustering with nested input objects

The Water Sensors task of Figure 8 is a composite task that contains three objects, the Water Sensors Coordinator, the High Water Sensor Interface and the Low Water Sensor Interface objects.

Connector classes encapsulate the details of inter-task communication, such as loosely and tightly coupled message communication. Some concurrent programming languages such as Ada and Java provide mechanisms for inter-task communication and synchronization. Neither of these languages supports loosely coupled message communication. In order to provide this capability, it is necessary to design a Message Queue connector class, which encapsulates a message queue and provides operations to access the queue.

A connector is designed using a monitor, which combines the concepts of information hiding and task synchronization [Bacon 2003, Magee and Kramer 2006]. These monitors are used in a single processor or multiprocessor system with shared memory. Connectors may be designed to handle asynchronous message communication, synchronous message communication without reply, and synchronous message communication with reply.

6. Software Architectural Patterns for Real-Time Control

Software architectural patterns [Buschmann 1996] provide the skeleton or template for the overall software architecture or high-level design of an application. Basing the software architecture of a product line on one or more software architectural patterns helps in designing the original architecture, because it is based on a proven architecture, as well as evolving the architecture.

There are two main categories of software architectural patterns [Gomaa 2011]. Architectural structure patterns address the static structure of the software architecture. Architectural communication patterns address the message communication among distributed components of the software architecture.

Most software systems can be based on well-understood overall software architectures. For example, the client/server software architecture is prevalent in many software applications. There is the basic client/server architecture, with one server and many clients. However, there are also many variations on this theme, such as multiple client/multiple server architectures and brokered client/server architectures.

Many real-time systems provide overall control of the environment by providing either centralized control, decentralized control, or hierarchical control. Each of these control approaches can be modeled using a software architectural pattern. In a centralized control pattern, there is one control component, which executes a statechart. It receives sensor input from input components and controls the external environment via output components, as shown in Figure 7 for the Pump Controller task.

In a centralized control pattern, the control component executes a statechart, which for Pump Controller is depicted in Figure 4. Another pattern used in the Pump Monitoring and Control System is the client/server pattern, as shown in Figure 5, where the Pump Subsystem is the client and the Pump Status Server is the server.

Architectural communication patterns for real-time systems include asynchronous communication and synchronous communication, both with and without reply. Other possible communication patterns include subscription/notification patterns and broker patterns. In the Pump Monitoring and Control System, both asynchronous and synchronous message communication are used as shown in Figures 5 and 7.

7. Performance Analysis of Real-Time Designs

Performance analysis of software designs is particularly important for real-time systems. The consequences of a real-time system failing to meet a deadline can be catastrophic.

The quantitative analysis of a real-time system design allows for the early detection of potential performance problems. The analysis is for the software design conceptually executing on a given hardware configuration with a given external workload applied to it. Early detection of potential performance problems allows different software designs and hardware configurations to be investigated.

In COMET, performance analysis of software designs is achieved by applying *real-time scheduling* theory. *Real-time scheduling* is an approach that is particularly appropriate for hard real-time systems that have deadlines that must be met [Gomaa 2000, SEI 1993]. With this approach, the real-time design is analyzed to determine whether it can meet its deadlines.

A second approach for analyzing the performance of a design is to use *event sequence analysis* and to integrate this with the *real-time scheduling* theory. Event sequence analysis considers scenarios of task collaborations and annotates them with the timing parameters for each of the tasks participating in each collaboration, in addition to system overhead for inter-object communication and context switching. The equivalent period for the active objects in the collaboration is the minimum inter-arrival time of the external event that initiates the collaboration.

8. Real-Time Embedded Software Product Line Design

A software product line (SPL) consists of a family of software systems that have some common functionality and some variable functionality [Parnas 1979, Clements 2002, Gomaa 2005]. Software product-line engineering involves developing the requirements, architecture, and component implementations for a family of systems, from which products (family members) are derived and configured. The problems of developing individual software systems are scaled upwards when developing software product lines because of the increased complexity due to variability management.

A better understanding of a system or product line can be obtained by considering the multiple views, such as requirements models, static models, and dynamic models of the system or product line. A graphical modeling language such as UML helps in developing, understanding, and communicating the different views.

A key view in the multiple views of a software product line is the feature modeling view. The feature model is crucial for managing variability and product derivation as it describes the product line requirements in terms of commonality and variability, as well as defining the product line dependencies. Furthermore, it is necessary to have a development approach that promotes software evolution, such that original development and subsequent maintenance are both treated using feature-driven evolution.

The Evolutionary Software Product Line Engineering Process [Gomaa 2005] is a highly iterative software process that eliminates the traditional distinction between software development and maintenance. Furthermore, because new software systems are outgrowths of existing ones, the process takes a software product line perspective; it consists of two main processes, as shown in Figure 9:

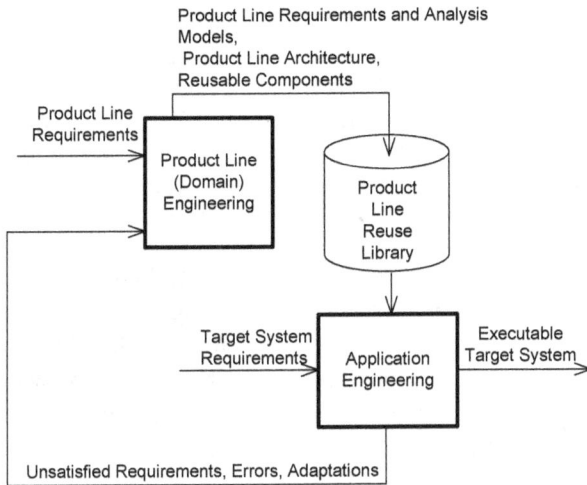

Figure 9: Process models for software product line engineering

- *Product Line (Domain) Engineering.* A product line multiple-view model, which addresses the multiple views of a software product line, is developed. The product line multiple-view model, product line architecture, and reusable components are developed and stored in the product-line reuse library.

- *Software Application Engineering.* A software application multiple-view model is an individual product line member derived from the software product line multiple-view model. The user selects the required features for the individual product line member. Given the features, the product line model and architecture are adapted and tailored to derive the application architecture. The architecture determines which of the reusable components are needed for configuring the executable application.

Software product-line concepts can also be applied to the design of embedded real-time software. Thus, the COMET design method has been extended to the

PLUS method (Product Line UML-Based Software engineering) for designing embedded real-time software product lines as described in [Gomaa 2005].

9. Conclusions

This paper has described concepts and methods for the design of concurrent and real-time software systems. When designing concurrent and real-time systems, it is essential to blend object-oriented concepts with the concepts of concurrent processing. This paper has given an overview of the COMET model-based software design method for designing concurrent and real-time systems, which integrates object-oriented and concurrent processing concepts and uses the UML notation.

For software-intensive systems, in which the software is one component of a larger hardware/software system, systems modeling can be carried out before software modeling. A dialect of UML called SysML is a general purpose modeling language for systems engineering applications [Friedenthal et al. 2009]. More information on UML modeling for real-time and embedded systems is given in MARTE, the UML profile for Modeling and Analysis of Real-Time and Embedded Systems [Espinoza et al. 2009].

With the proliferation of low cost workstations and personal computers operating in a networked environment, the interest in designing concurrent systems, particularly real-time and distributed systems, is growing rapidly. Furthermore, with the growing need for reusable designs, design methods for software product lines [Gomaa 2005] and service-oriented architectures [Gomaa 2011] are likely to be of increasing importance for future real-time embedded software systems.

REFERENCES

- Awad, M., J. Kuusela, and J. Ziegler, *Object-Oriented Technology for Real-Time Systems,* Prentice Hall, NJ, 1996.

- Bacon, J., *Concurrent Systems*, 3rd ed., Addison-Wesley, Reading, MA, 2003.

- Booch, G. 1994, *Object-Oriented Analysis and Design with Applications,* 2nd ed., Addison-Wesley Professional, Boston, 1994.

- Booch, G., R. A. Maksimchuk, M. W. Engel, et al., *Object-Oriented Analysis and Design with Applications,* 3rd ed., Addison-Wesley, Boston, 2007.

- Booch, G., J. Rumbaugh, I. Jacobson, *The Unified Modeling Language User Guide,* 2nd ed. Addison-Wesley, Reading, MA, 2005.

- Buhr, R.J.A. and R.S. Casselman, *Use Case Maps for Object-Oriented Systems,* Prentice Hall, NJ, 1996.

- Buschmann, F., R. Meunier, H. Rohnert, and P. Sommerlad, *Pattern Oriented Software Architecture: A System of Patterns,* John Wiley & Sons, NY, 1996.

- Clements, P. and L. Northrop, *Software Product Lines: Practices and Patterns,* Addison-Wesley, Reading, MA, 2002.

- Coad, P. and E. Yourdon, *Object-Oriented Analysis,* Prentice Hall, NY, 1991.

- Douglass, B. P., *Doing Hard Time: UML, Objects, Frameworks, and Patterns in Real-Time Software Development,* Addison-Wesley, Reading, MA, 1999.

- Douglass, B. P., *Real-Time UMeL,* 3rd, Addison-Wesley, Reading, MA, 2004.

- Espinoza H., D. Cancila, B. Selic and S. Gérard "Challenges in Combining SysM and MARTE for Model-Based Design of Embedded Systems," *Lecture Notes in Computer Science,* Vol. 5562, pp. 98-113. Springer, Berlin, 2009.

- Fowler, M. and K. Scott, *UML Distilled,* 3rd ed., Addison-Wesley, Reading, MA, 2004.

- Friedenthal, S., A. Moore, and R. Steiner, *A Practical Guide to SysML: The Systems Modeling Language,* Morgan Kaufmann, Burlington, MA, 2009.

- Gomaa, H., "A Software Design Method for Real Time Systems, *Communications of the ACM,* Vol. 27, No. 9, September 1984.

- Gomaa, H., *Software Design Methods for Concurrent and Real-Time Systems,* Addison-Wesley, Reading, MA, 1993.

- Gomaa, H., *Designing Concurrent, Distributed, and Real-Time Applications with UML,* Addison-Wesley, Reading, MA, 2000.

- Gomaa, H., Gomaa, H., and M.E. Shin, "A Multiple-View Meta-Modeling Approach for Variability Management in Software Product Lines, Proc., *International Conference on Software Reuse,* Madrid, Spain, Springer LNCS 3107, July 2004.

- Gomaa, H., *Designing Software Product Lines with UML: From Use Cases to Pattern-based Software Architectures,* Addison-Wesley, Reading, MA, 2005.

- Gomaa, H., "A Software Modeling Odyssey: Designing Evolutionary Architecture-Centric Real-Time Systems and Product Lines," Keynote paper, *Proc. ACM/IEEE 9th International Conference on Model-Driven Engineering, Languages and Systems,* Springer Verlag LNCS 4199, Pages 1-15, Genova, Italy, October 2006.

- Gomaa, H., *Software Modeling, and Design: UML, Use Cases, Patterns, and Software Architectures*, Cambridge University Press, NY, 2011.

- Gomaa, H. and M.E. Shin, "A Multiple-View Meta-Modeling Approach for Variability Management in Software Product Lines," *Proc. International Conference on Software Reuse*, Madrid, Spain, Springer LNCS 3107, July 2004.

- Harel, D., "On Visual Formalisms," *CACM* 31, 5 (May 1988), 514-530.

- Harel, D. and E. Gary, "Executable Object Modeling with Statecharts," *Proc. 18th International Conference on Software Engineering*, Berlin, March 1996.

- Harel, D. and M. Politi, *Modeling Reactive Systems with Statecharts*, McGraw-Hill, 1998.

- Hatley D. and I. Pirbhai, *Strategies for Real Time System Specification*, Dorset House, NY, 1988.

- Jackson, M., *System Development*, Prentice Hall, NJ, 1983.

- Jacobson, *Object-Oriented Software Engineering*, Addison-Wesley, Reading, MA, 1992.

- Jacobson, G. Booch, J. Rumbaugh, *The Unified Software Development Process,* Addison-Wesley, Reading, MA, 2000.

- Magee, J. and J. Kramer, *Concurrency: State Models & Java Programs,* John Wiley & Sons, 2nd ed, 2006.

- Parnas, D., "Designing Software for Ease of Extension and Contraction," *IEEE Transactions on Software Engineering*, March 1979.

- Parnas, D., P. Clements and D. Weiss, "The Modular Structure of Complex Systems," *Proc. Seventh IEEE International Conference on Software Engineering*, Orlando, Florida, March 1984.

- Rosenberg, D., and K. Scott, *Use-Case Driven Object Modeling with UML*, Addison-Wesley, Reading, MA, 1999

- Rumbaugh, j., J. Blaha, W. Premerlani, F. Eddy, W. Lorenson, *Object-Oriented Modeling and Design*, Prentice Hall, NJ, 1991.

- Rumbaugh, J., G. Booch, I. Jacobson, *The Unified Modeling Language Reference Manual*, 2nd ed., Addison-Wesley, Reading, MA, 2005.

- SEI (Carnegie Mellon, Software Engineering Institute), *A Practioner's Handbook for Real-Time Analysis - Guide to Rate Monotonic Analysis for Real-Time Systems,* Kluwer Academic Publishers, Boston, 1993.

- Selic, B., G. Gullekson, and P. *Ward, Real-Time Object-Oriented Modeling*, Wiley & Sons, NY, 1994.

- Shaw, M. and D. Garlan, *Software Architecture: Perspectives on an Emerging Discipline*, Prentice Hall, NJ, 1996.

- Simpson, H. and K. Jackson, "Process Synchronization in MASCOT," *The Computer Journal*, Vol.17, no. 4, 1979.

- Simpson, H., "The MASCOT Method," *IEE/BCS Software Engineering Journal*, 1(3), 1986, 103-120.

- Shlaer, S. and S. Mellor, *Object Oriented Systems Analysis*, Prentice Hall, NJ, 1988.

- Silberschatz, A., P. Galvin, and G. Gagne, *Operating System Concepts*, 8th ed., Addison-Wesley, Reading, MA, 2008.

- Tanenbaum A.S, *Modern Operating Systems*, 3rd ed., Prentice Hall, NJ, 2008.

- Taylor, R.N., N. Medvidovic, E.M. Dashofy, *Software Architecture: Foundations, Theory, and Practice*, Wiley & Sons. NY, 2009.

- Ward, P. and S. Mellor, *Structured Development for Real-Time Systems*, Vols. 1, 2 & 3, Yourdon Press, NY,1985.

- Yourdon, E., *Modern Structured Analysis*, Prentice Hall, NJ, 1989.

Chapter 4
Software Design Documentation Standard[3]

Abstract

Engineering documentation standards are documented agreements containing technical specifications or other precise criteria to be used consistently as engineering processes, rules, guidelines, and definitions of characteristics, to ensure that materials, products, and services are fit for their purpose.

Standards provide rules, guide lines and heuristics that, if followed, deliver an assurance of good practice—they are not intended to be about best practice.

The results of using this standard is a complete and reasonably correct software design description (SDD). The SDD is a description of a software system to be developed. It lays out architectural and detailed design descriptions.

The software design description implements the software requirements specification establishing the basis for an agreement between customers and contractors or suppliers (in market-driven projects, these roles may be played by the marketing and development divisions) detailing what the software code is expected to do as well as what it is not expected to do. Software design documentations permits a rigorous implementation of the project requirements prior to beginning software construction (coding). It should also help in estimating product costs, risks, and schedules.

The purpose of this document is to (1) aid the IEEE Software Requirements Certificate of Proficiency exam candidates in preparing, taking and passing the Software Design exam, (2) provide an example of a software design decumation standard to demonstrate the usefulness of an IEEE software engineering standard, and (3) provide a template for university students to use when writing an SDD for a classroom software engineering project.

---

3. This article is modeled after *IEEE Standard 1016-1998*, "IEEE Recommended Practice for Software Design Descriptions." This standard has been superseded by IEEE Standard 1016-2009. Also, this version is incomplete and a simplified copy of *Standard 1016-1998*. This version of the standard cannot be used to cite adherence to an IEEE standard in a contractual situation. Its purpose is to provide the student and the test taker with a basic understanding and use of IEEE software engineer standards.

INTRODUCTION

The necessary information content and recommendations for an organization for Software Design Descriptions (SDDs) are described. An SDD is a representation of a software system that is used as a medium for communicating software design information. This recommended practice is applicable to paper documents, automated databases, design description languages, or other means of description.

1. Outline of a Software Design Description

This section discusses each of the essential parts of the SDD.

1.1 Introduction.

This section of the SDD should state the purpose and the scope of the SDD, reference materials, and definitions and acronyms used within the SDD. This section is provided to aid the reader in understanding the design document.

1.1.1 Purpose of the SDD. This subsection should:

(1) *Identify the document.* For example, the specification could start with, This document establishes the design of a software system identified as . . . (insert name of software system).

(2) *Delineate the specific purpose* of the particular SDD.

(3) *Specify the intended audience* for the SDD.

1.1.2 Scope of the SDD. The subsection should:

(1) Identify the software product(s) to be produced by name. For example: Host DBMS, Report Generator, AJAX Parole Program, F-1 Fire Control System, etc.

(2) Explain what the software product(s) will, and, if necessary, will not do.

(3) Describe the application of the software being specified. As a portion of this, it should:

   a. Describe all relevant benefits, objectives, and goals as precisely as possible. For example, to say that a goal is to provide effective reporting capabilities would be better said as the goal is a parameter-driven, user-definable reports with a two-hour turnaround and on-line entry of user parameters.

   b. Be consistent with similar statements in higher-level specifications (system/software requirements specifications, software contract, and statement of work), if they exist.

1.1.3 Definitions, acronyms, and abbreviations. This subsection should provide the definitions of all terms, acronyms, and abbreviations required to properly interpret the SDD. This information may be proved by reference to one or more appendices in the SDD or by reference to other documents.

1.1.4 References. This subsection should:

(1) Provide a complete list of all documents referenced elsewhere in the SDD, or in a separate, specified document.

(2) Identify each document by title; report number, if applicable; date; and publishing organization.

(3) Specify the sources from which the references can be obtained. This information may be provided by reference to an appendix or to another document.

1.1.5 Overview. This subsection should:

(1) Describe what the remainder of the SDD contains. It is not necessary to describe Section 1 in this overview; since this is the end of Section 1 (we can presume the reader already knows what is included in it).

(2) Explain how the SDD is organized. Be sure to include an explanation of the appendices if they are considered to be a formal part of the SDD.

2. General Design Considerations

This section of the SDD describes the general design of the system and provides assumptions, constraints, and decisions which affect the entire system. The purpose of this section is to provide information, which makes the specific design in the following sections.

2.1 Overview of the system.

This subsection should contain some general considerations that will help in formulating the design considerations.

2.1.1 Assumptions. This subsection should describe all design assumptions made by the designers. An assumption is something that is taken to be true because we are either unable or unwilling to determine the actual truth.

While it is very difficult to isolate all of the assumptions, which one makes during design, a significant effort should be made in that direction. Many projects have failed because assumptions were made unconsciously, assumptions which later proved to be incorrect. By documenting assumptions in this subsection, management and other readers will at least be aware of the assumptions that the designers made. Potentially dangerous assumptions may be uncovered, which will save many man-hours of work later. At the very least, the designers will have acted to cover themselves should one of their assumptions prove false.

Some typical assumptions which need to be assessed involve availability and quality of other software and hardware (and particularly, availability of the documentation for that software and hardware) and technical assumptions for projects which have facets that are unlike anything which the design team has dealt with in the past. An example from a recent senior project involved the assumption that a Terminate-and-Stay-Resident (TSR) program on a microcomputer using MS-DOS would be able to make calls on the operating system without any special considerations. This assumption proved to be false and the proposed system was never completed.

2.1.2 Constraints. This subsection should describe any additional constraints on the way this system must be designed and implemented. It is the intent of the SDD to carry out the requirements expressed in the SRS. The SRS, therefore, is the primary constraint upon the design. Clearly, it is unnecessary to restate the requirements from the SRS here as constraints. However, it is likely that certain requirements impose additional constraints on the design. For example, performance requirements may constrain the types of algorithms which may be used or hardware considerations may constrain data representations (files versus arrays). The choice of an implementation language may impose further constraints, as may the use of other software products.

2.1.3 Major decisions. One of the most important aspects of an SDD is a discussion of the decisions made by the designers. Decisions which affect only a specific element should be discussed in the specific design subsection. However, any decisions which affect the entire design or a substantial portion of the design should be explained here. This information is invaluable to the maintenance function and also allows for easy access to those decisions during reviews.

Decisions on shared data structures, language features to be used, argument passing versus global data, and module size are only a few of the aspects that may require discussion in this subsection.

2.1.4 Conventions. In this subsection, the designers should explain all special notations and other conventions which will be used in the specific design. In addition, there is usually a body of general advice which the designers wish to provide for the programmers. Rather than repeating this advice in each element description, it may be collected here and made available to all of the programmers.

3. Architectural Design Considerations

This subsection should be aimed at management-level readers of the SDD. It should provide sufficient depth to allow these readers to determine whether or not the design fulfills the general requirements as established in the SRS.

Software architecture refers to the high level structures of a software system, the discipline of creating such structures, and the documentation of these structures. These structures are needed to reason through the software system. Each structure comprises software elements, relations among them, and properties of both elements and relations.

Software architecture is about making fundamental structural choices, which are costly to change once implemented. Software architecture choices include specific structural options from possibilities in the design of software. For example, the systems that controlled the Space Shuttle launch vehicle had the requirements of being very fast and very reliable. Therefore, an appropriate real-time computing language would need to be chosen. Additionally, to satisfy the need for reliability, the choice could be made to have multiple redundant and independently produced copies of the program, and to run these copies on independent hardware while cross-checking results.

Each software engineering system is comprised of separate, semi-autonomous components, called configuration items (CI). This collection of configuration items is termed the architecture of the systems. Each CI is each comprised of software design entries—typically called modules. Each module is named and described independently. (*See Paragraph 4*). Interactions between configuration items and modules need to be kept to a minimum (*See Paragraph 2.1.4*).

Each CI and accompanying module needs to be traced to the software requirement that it executes, which in turn needs to be traced to the code that it implements. These tracings need to be documented in a traceability matrix—either manual or automated.

Each CI gave rise to a number of interesting ideas about SED at different levels of abstraction. Some of these concepts can be useful during the architectural design of specific software (for example, architectural style), as well as during its detailed design (for example, lower-level design patterns). However, they can also be useful for designing generic systems, leading to the design of families of programs (also known as product lines). Interestingly, most of these concepts can be seen as attempts to describe, and thus reuse generic design knowledge [SWEBOK 2004].

This subsection should describe the top-level components (design elements) of the system. The names of these elements should be consistent with the names used in the identification attribute. Each major element should be described in sufficient detail to allow the readers of the SDD to comprehend the more detailed descriptions to follow in the specific design. Particular emphasis should be placed on explaining the interrelationships between the elements.

Sommerville [2011] has identified a number of decisions that must be made during the design process by the system architects. They have to consider the following fundamental questions about the system:

(1) Is there a generic application architecture that can act as a template for the system that is being designed?

(2) How will the system be distributed across a number of cores or processors?

(3) What architectural patterns or styles might be used?

(4) What fundamental approach will be used to structure the system?

(5) How will the structural components in the system be decomposed into subcomponents?

(6) What strategy will be used to control the operation of the components in the system?

(7) What architectural organization is best for delivering the nonfunctional requirements of the system?

(8) How will the architectural design be evaluated?

(9) How should the architecture of the system be documented?

4. Detailed Design Considerations

4.1 Detailed design modules.

This section of the SDD should contain the descriptions of all attributes for each design element, i.e., modules. This is typically the largest part of the SDD.

A software detailed design is a representation or model of the software system to be created. The model should provide the precise design information needed for planning, analysis, and implementation of the software system. It should represent a partitioning of the system into design entities and describe the important properties and relationships among those entities.

The design description model used to represent a software system can be expressed as a collection of design entities, each possessing properties and relationships. To simplify the model, the properties and relationships of each design entity are described by a standard set of attributes. The design information needs of project members are satisfied through identification of the entities and their associated attributes. A design description is complete when the attributes have been specified for all the entities [IEEE Std. 1016-1998].

4.2 Design entities.

A *design entity* (a.k.a. a *design module*) is an element (component) of a design that is structurally and functionally distinct from other elements and that is separately named and referenced. Design entities result from a decomposition of the software system requirements. The objective is to divide the system into separate components that can be considered, implemented, changed, and tested with minimal effect on other entities. [IEEE Std. 1016-1998].

The number and type of entities required to partition a design are dependent on a number of factors, such as the complexity of the system, the design technique used, and the programming environment.

Although entities are different in nature, they possess common characteristics. Each design entity will have a name, purpose, and function. There are common relationships among entities such as interfaces or shared data. The common characteristics of entities are described by design entity attributes.

4.3 Design entity attributes.

A *design entity attribute* is a named characteristic or property of a design entity. It provides a statement of fact about the entity.

Design entity attributes can be thought of as questions about design entities. The answers to those questions are the values of the attributes. All the questions can be answered, but the content of the answer will depend upon the nature of the entity. The collection of answers provides a complete description of an entity.

The list of design entity attributes presented in this sub clause is the minimum set required for all SDDs.

All attributes shall be specified for each entity. Attribute descriptions should include references and design considerations such as tradeoffs and assumptions when appropriate. In some cases, attribute descriptions may have the value *none*. When additional attributes are identified for a specific software project, they should be included in the design description.

The attributes and associated information items are defined in Sections 4.3.1 through 4.3.12.

4.3.1 Identification. *The name of the entity.* Two entities shall not have the same name. The names for the entities may be selected to characterize their nature. This will simplify referencing and tracking in addition to providing identification for each entity.

4.3.2 Designer. *The name of the designer.* This attribute will list the names of the designers of the design element, their organizations, and points of contact. It will provide a resource for additional background information not included in the description.

4.3.3 Type. *The nature of the element.* The type attribute shall describe the nature of the element such as subprogram, module, procedure, package, etc.

4.3.4 Source. *The source of the element.* This attribute will list the source of the element.

The software for the design element may be developed internally, reused from an earlier project or release, or acquired from an outside source. The source of the software should be supplied to aid in budgeting, staffing, and schedule estimation.

43.5 Purpose. *Purpose identifies the requirement that this element satisfies.* This subsection of the SDD should identify the requirements from the SRS that the element satisfies. If the element does not directly satisfy an SRS requirement, then this subsection should identify the superordinate elements that use this element. In identifying the superordinate elements, a description of the rationale used for creating this element should be given. This information is needed to provide a clear transformation from requirements specification to design specification.

4.3.6 Function. *A statement of what the entity does.* The function attribute shall state the transformation applied by the entity to inputs needed to produce the desired output. In the case of a data entity, this attribute shall state the type of information stored or transmitted by the entity.

4.3.7 Subordinates. *The identification of all entities composing this entity.* The subordinates attribute shall identify the *composed of* relationship for an entity. This information is used to trace requirements to design entities and to identify parent/child structural relationships through a software system decomposition.

4.3.8 Dependencies. *A dependency is a description of the relationships of this entity with other entities.* The dependencies attribute shall identify the *uses* or *requires the presence of* relationship for an entity. These relationships are often graphically depicted by structure charts, data flow diagrams, and transaction diagrams.

This attribute shall describe the nature of each interaction including such characteristics as timing and conditions for interaction. The interactions may involve the initiation, order of execution, data sharing, creation, duplicating, usage, storage, or destruction of entities.

4.3.9 Interface. *Interface provides a description of how other entities interact with this entity.* The interface attribute shall describe the *methods* of interaction and the *rules* governing those interactions. The methods of interaction include the mechanisms for invoking or interrupting the entity, for communicating through parameters, common data areas or messages, and for direct access to internal data. The rules governing the interaction include the communications protocol, data format, acceptable values, and the meaning of each value.

This attribute shall provide a description of the input ranges, the meaning of inputs and outputs, the type and format of each input or output, and output error codes. For information systems, it should include inputs, screen formats, and a complete description of the interactive language.

4.3.10 Resources. *Resources* describe *the elements used by the entity that are external to the design.* The resources attribute shall identify and describe all of the resources *external* to the design that are needed by this entity to perform its function. The interaction rules and methods for using the resource shall be specified by this attribute.

This attribute provides information about items such as physical devices (printers, disc-partitions, memory banks), software services (math libraries, operating system services), and processing resources (CPU cycles, memory allocation, buffers).

The resources attribute shall describe usage characteristics such as the process time at which resources are to be acquired and sizing to include quantity, and physical sizes of buffer usage. It should also include the identification of potential race and deadlock conditions as well as resource management facilities.

4.3.11 Processing. *Processing is a description of the rules used by the entity to achieve its function.* The processing attribute shall describe the algorithm used by the entity to perform a specific task and shall include contingencies. This description is a refinement of the function attribute. It is the most detailed level of refinement for this entity.

This description should include timing, sequencing of events or processes, prerequisites for process initiation, priority of events, processing level, actual process steps, path conditions, and loop back or loop termination criteria. The handling of contingencies should describe the action to be taken in the case of overflow conditions or in the case of a validation check failure.

4.3.12 Data. *A description of data elements internal to the entity.* The data attribute shall describe the method of representation, initial values, use, semantics, format, and acceptable values of internal data.

The description of data may be in the form of a data dictionary that describes the content, structure, and use of all data elements. Data information shall describe everything pertaining to the use of data or internal data structures by this entity. It shall include data specifications such as formats, number of elements, and initial values. It shall also include the structures to be used for representing data such as file structures, arrays, stacks, queues, and memory partitions.

The meaning and use of data elements shall be specified. This description includes such things as static versus dynamic, whether it is to be shared by transactions, used as a control parameter, or used as a value, loop iteration count, pointer, or a link field. In addition, data information shall include a description of data validation needed for the process.

Chapter 5

Software Design Exercises

These design entity exercises are provided to encourage you to browse the chapter looking for answers to the questions provided. If truth be told, the correct answer for all software engineering questions is it depends. To avoid this issue, a set of possible answers is provided. There is (supposedly) only one correct answer.

If you are using this book as a textbook in a university course, your instructor may require you to justify your answer. (Why are some of the possible answers correct and why are some of them wrong?) The instructor might also ask you to identify any assumptions you depended on when arriving at your answer.

However, if you are very clever, maybe you can come up with more than one correct answer (which of course you would have to justify).

1.   **Which of the following is NOT a valid use of a baseline?**

   [a] To distinguish between different internal releases for delivery to a customer
   [b] To help ensure complete and up-to-date documentation
   [c] To enforce standards
   [d] To control changes to the executable code modules

2.   **A benefit of human computer interface (HCI) is:**

   I. Reduced learning times
   II. Lower error rates
   III. The individual skill of the user can be ignored
   IV. All like processes are on one screen

   [a] I and II
   [b] I and III
   [c] I and IV
   [d] II and IV

3.   **The following diagrams are NOT created in performing object-oriented design:**

   [a] Class
   [b] Activity
   [c] Use-case
   [d] Sequence

4. A work breakdown structure is:

    I.   A means of representing a product
    II. A means of representing a process
    III. Required in developing top-down software costs
    IV. Only used in embedded computer systems

    [a] I only
    [b] I and II
    [c] I, II, and III
    [d] II and IV

6. The degree of interaction between two modules is known as:

    [a] Cohesion
    [b] Strength
    [c] Inheritance
    [d] Coupling

6. An object identified during object-oriented analysis:

    [a] Is a design object
    [b] Has no relationship with any object identified during subsequent design
    [c] May have a mapping to one or more objects identified during subsequent design
    [d] Has no attributes identified for it

7. How is peer defined in peer reviews?

    I.   Peers are equals of the product's author.
    II. Peers are highly educated individuals who are used to review document for errors.
    III. Peers are technical gurus who are used to analyze documents, looking for errors.
    IV. Peers have no undue influence over the authors' careers or pay scale.

    [a] I and IV
    [b] II only
    [c] III only
    [d] IV only

8. All of the following are major aspects of a software design EXCEPT:

[a] Control structures
[b] Algorithms
[c] Requirements
[d] Requirements traceability matrix

9. When planning incremental builds, which of the following should be considered by a project manager?

I. Can the requirements be completed prior to beginning the first build?
II. Can the software system be portioned into subsystems?
III. Are corresponding regression tests available?
IV. Can the subsystems be tested independently?

[a] I and II
[b] I, II, and III
[c] II and III
[d] I, III, and IV

10. The input to the design process is the:

[a] Requirements specification document
[b] ConOps document
[c] Test case document
[d] Software engineering management plan

11. The following are all statements regarding possible properties of a user-interface:

I. It provides increased confidence in the software.
II. It causes a reduction in errors made by the user.
III. It results in increased customer satisfaction.
IV. It leads to greater efficiency in accomplishing the task.

Which set of these statements do you regard as being true?

[a] I only
[b] I and II
[c] I, II, and III
[d] I, II, III, and IV

12. **Which of the following notations describes the static view of a software design?**

[a] Class diagrams, sequence diagrams and component diagrams
[b] Sequence diagrams, class diagrams and collaboration responsibility cards (CRCs)
[c] State transition diagrams, sequence diagrams, and component diagrams
[d] Class diagrams, component diagrams, and sequence diagrams

13. **Which of the following source code metrics could be evaluated using a static analysis tool? You may assume that the source code successfully compiles.**

I. Number of faults in the source code
II. Distribution of statement types in the source code
III. Density of comments in the source code
IV. Appropriateness of comments in the source code

[a] I and II
[b] III and IV
[c] II and III
[d] I and IV

14. **The following are typical properties of software systems for solving real-world problems:**

I. The software is invisible.
II. The software is complex.
III. The software is large.
IV. The software is easily changeable.

Which of the above properties makes software design particularly difficult?

[a] I, II, and III
[b] I, II, and IV
[c] I, III, and IV
[d] II, III, and IV

15. **A good design will accomplish all of the following EXCEPT:**

   [a] Implement all explicit and implicit requirements
   [b] Provide guidance for software requirements engineers
   [c] Address data, functional, and behavioral domains from an implementation perspective
   [d] Set up a requirements traceability matrix

16. **Which one of the following is NOT a recognized software design method?**

   [a] Divide & conquer design
   [b] Structured design
   [c] Object oriented design
   [d] Component based design

17. **The main advantage of structured programming is:**

   [a] It is more efficient
   [b] It tends to be more reliable
   [c] It is easier to write
   [d] It can be flowcharted

18. **The elements of the software architecture of a computing system include:**

   I.   **Software components**
   II. **Class diagrams**
   III. **Connectors expressing relationships between software compoents**
   IV. **Entity-relationship diagrams**

   [a] I and II
   [b] I and III
   [c] I, III, and IV
   [d] I, II, III, and IV

19.   The elements of the software architecture of a computing system in-
        clude:

        I.   Software component
        II. Class diagrams
        III. Connectors expressing relationships between components
        IV. Entity-relationship diagrams

        [a] I and II
        [b] I and III
        [c] II only
        [d] IV only

20.   The following are all statements regarding possible properties of a
        user-interface.

        I.   It provides increased confidence in the software
        II. It causes a reduction in errors made by the user
        III. It results in increased customer satisfaction
        IV. It leads to greater efficiency in accomplishing the task

        Which set of these statements do you regard as being true?

        [a] I only
        [b] I & II
        [c] I, II, & III
        [d] I, II, III, & IV

INDEX

abstraction, 6, 40, 51
agile manifesto, 73, 75
agile methods, 56, 68, 70, 72, 73, 74
architectural design, 2, 3, 6, 12, 14-16, 32, 48, 107, 108
architectural style, 12, 13, 45, 58, 59, 61, 64, 66, 69, 107
architecture description languages, 32

batch architecture, 15, 50
Bauer, Friedrich I, viii, x
behavioral descriptions (dynamic view), 33
behavioral design patterns, 17
behavioral patterns, 71
binding, 7, 8
blackboard architecture, 13, 46
Boehm, Barry, 72
broker architecture, 14, 48
Budgen, David, iii, v, 53

client-server architecture, 14, 46
cohesion, 7, 114
color, 24
COMET method, 84
component-based development, 74
concerns, separation of, 9
concurrency, 3, 9, 10, 82, 100
concurrent, 9, 10, 42, 46, 81, 82, 83, 84, 85, 86, 89, 91, 93, 95, 98
concurrent processing, 81, 82, 98
constructional viewpoint, 64
coupling, 7, 114
creational design patterns, 16
creational patterns, 71

data, 3, 7, 10, 21, 33, 40, 41, 44, 59, 65, 68, 110, 111
data flow diagram (DFD), 35-37
data persistence, 3, 10
data-structure-centered design, 2, 41
deadlock, 9
decision tables, 33, 44
decomposition, 8
dependencies, 20, 110

design, i, ii, iii, v, ix, x, 1,-109 passim
design assumptions, 105
design concepts, 3, 4
design entity attribute, 20, 108
design entity exercises, 113
design is a wicked problem, 4
detailed design, 2, 4, 612, 19, 32, 33, 94, 103, 107, 108
diagrams, 20, 32, 33, 36, 37, 41, 60, 62, 63, 68, 70, 76, 83-85, 91, 93, 110-117, passim
divide and conquer, 34
documentation standards, 103
dynamic modeling, 83, 84, 89

encapsulation, 8, 40
entity-relationship diagram, 66
entity-relationship model, 44
error handling, 10
event-driven interface, 26
event-driven system, 10
events, 2, 3, 10, 21, 26, 82, 86, 90, 110
exception handling, 2, 3, 10, 11
extreme programming (XP), 73

fan-in/fan-out, 38
fault tolerance, 10
feedback, 3, 23, 25, 27, 80
fonts, 24, 25
function, viii, 3, 20, 36, 109
functional viewpoint, 62, 65

Garlan, David, 51, 57, 58
Gomaa, Hassan, iii, v, 81

Harel, David, 67, 78, 83, 89, 100
HCI design, 3, 22, 31
HCI, psychology of, 3, 31
heuristics, 35
HIPO diagrams, 44
HTTP, 30, 42, 43, 44
HTTPS, 43
human-computer interface (HCI), 22

icons, 25
identification, 20, 109

IEEE standard 1016-1998, x, 103
information hiding, 6, 8, 33, 34, 83, 95
inheritance, 40, 114
input/output, 28, 29
interaction, 3, 11, 91
interface, v, 1, 3, 21, 22, 23, 27, 32, 89, 94, 110, 113, 117
interpreter architecture, 15, 50

knowledge schema, 55, 67, 68, 77
Kruchten, Phillipe, 61, 79

languages, 29
layered architecture, 13, 45

mathematical notations, 63
measures, 3, 7
menu-driven interface, 26
metaphors, 3, 30
Microkernel architecture, 14, 49
model-view-controller (MVC) architecture, 14, 48
modes, use of, 3, 23
modularization, 8
multimedia, 3, 28

navigation, use of, 3, 23

object structuring criteria, 85, 86, 88
object-oriented design, 3, 40
object-oriented software life cycle, 85
observer, 13, 46, 71, 72

pair programming, 73
pattern, 13, 14, 15, 16, 17, 18, 34, 35, 47, 48, 49, 58, 59, 60, 70, 71, 72, 96
plan-driven design, 68
polymorphism, 40
presentation-abstract-controller (PAC) architecture, 14, 48
process control architecture, 15, 50
processing, v, 21, 29, 82, 110
pseudo-code, 44
pump monitoring and control system class context diagram, 88

quality attributes, 3
quality factors, 56

real-time embedded systems, 84
real-time systems, 81, 82, 83, 84, 86, 93, 96, 98

reflection architecture, 15, 49
resources, 21, 110
response time, 3, 25
responsibility-driven design, 40, 41
rule of thumb, 37, 62
rule-based architecture, 15, 51
RUP, 42, 43

SADT, 42, 43
scope of effect versus scope of control, 38
Scrum, 73, 74, 79
security, 3, 11, 43
separation of concerns, 9
Shaw, Mary, 51, 58
SOAP, 42, 43, 75
software architecture, v, 6, 11, 12, 32, 36, 51, 57, 58, 78-80, 85, 91, 95, 96, 101, 106
software construction, 5
software design, ix, 1, 2, 3, 4, 5, 6, 37, 44, 54, 76, 103
software design knowledge areas, 3
software engineering design, vi, 1, 4
software product line (SPL), 97
software product line engineering, 97
software requirements, vi, 5
software systems, family of, 97
software testing, 5
software tools, 76
Sommerville, Ian, 13, 16, 22, 24, 45, 46, 48, 71, 79, 107
sound, 29
spiral model, 72, 77
starvation, 9
state dependent dynamic analysis, 89
statechart, 4367, 70, 89, 91, 96
stovepipe architecture, 13, 46
structural design patterns, 17
structural patterns, 71
structure charts, 32, 36-38, 44, 45, 64
structured English, 45

task structuring, 83, 84, 85, 86, 93
test-first programming, 73
text, 30, 62
three-tier architecture, 14, 47
top-down versus bottom-up strategies, 34
transaction analysis, 36, 68
transform analysis, 36, 68
transition, 33, 50, 83, 91, 115

UML activity diagram, 65
unified modeling language (UML), 64, 81, 84

VDM, 42
Vienna development method (VDM), 42
viewpoints, 2, 3, 12, 61, 62, 63, 64

waterfall model, 56, 72
webbed topology, 23
wicked problem, 4, 54

NOTES

NOTES

NOTES

NOTES

www.ingramcontent.com/pod-product-compliance
Lightning Source LLC
Chambersburg PA
CBHW080558220326
41599CB00032B/6525